Optical Signal Processing in Highly Nonlinear Fibers

Optical Signal Processing in Highly Nonlinear Fibers

Mário F. S. Ferreira

CRC Press
Taylor & Francis Group
Boca Raton London New York

CRC Press is an imprint of the
Taylor & Francis Group, an **informa** business

First edition published 2020
by CRC Press
6000 Broken Sound Parkway NW, Suite 300, Boca Raton, FL 33487-2742

and by CRC Press
2 Park Square, Milton Park, Abingdon, Oxon, OX14 4RN

© 2020 Taylor & Francis Group, LLC

CRC Press is an imprint of Taylor & Francis Group, LLC

Reasonable efforts have been made to publish reliable data and information, but the author and publisher cannot assume responsibility for the validity of all materials or the consequences of their use. The authors and publishers have attempted to trace the copyright holders of all material reproduced in this publication and apologize to copyright holders if permission to publish in this form has not been obtained. If any copyright material has not been acknowledged please write and let us know so we may rectify in any future reprint.

Except as permitted under U.S. Copyright Law, no part of this book may be reprinted, reproduced, transmitted, or utilized in any form by any electronic, mechanical, or other means, now known or hereafter invented, including photocopying, microfilming, and recording, or in any information storage or retrieval system, without written permission from the publishers.

For permission to photocopy or use material electronically from this work, access www.copyright.com or contact the Copyright Clearance Center, Inc. (CCC), 222 Rosewood Drive, Danvers, MA 01923, 978-750-8400. For works that are not available on CCC please contact mpkbookspermissions@tandf.co.uk

Trademark notice: Product or corporate names may be trademarks or registered trademarks, and are used only for identification and explanation without intent to infringe.

Library of Congress Cataloging-in-Publication Data

Names: Ferreira, Mário F. S., author.
Title: Optical signal processing in highly nonlinear fibers / Mário Ferreira.
Description: First edition. | Boca Raton, FL : CRC Press, 2020. | Includes bibliographical references and index. | Summary: "This book provides an updated description of the most relevant types of highly nonlinear fibers. It also describes some of their actual applications for nonlinear optical signal processing"-- Provided by publisher.
Identifiers: LCCN 2020004393 | ISBN 9780367205409 (hardback) | ISBN 9780429262111 (ebook)
Subjects: LCSH: Fiber optic cables. | Signal processing. | Optical data processing.
Classification: LCC TK5103.15 .F47 2020 | DDC 621.36/92--dc23
LC record available at https://lccn.loc.gov/2020004393

ISBN: 978-0-367-20540-9 (hbk)
ISBN: 978-0-429-26211-1 (ebk)

Typeset in Times
by Lumina Datamatics Limited

Contents

Author ix

Chapter 1 Introduction 1

Chapter 2 Nonlinear Effects in Optical Fibers 5

    2.1   Introduction 5
    2.2   The Kerr Effect 5
    2.3   The Nonlinear Schrödinger Equation 6
    2.4   Self-Phase Modulation 8
    2.5   Cross-Phase Modulation 10
    2.6   Four-Wave Mixing 11
    2.7   Stimulated Raman Scattering 15
    2.8   Stimulated Brillouin Scattering 18
    References 20

Chapter 3 Optical Solitons 25

    3.1   Introduction 25
    3.2   Soliton Solutions of the Nonlinear Schrödinger Equation 26
    3.3   Perturbations of Solitons 28
        3.3.1   Fiber Losses 28
        3.3.2   Higher-Order Effects 29
        3.3.3   Timing Jitter 30
    3.4   Soliton Transmission Control 31
        3.4.1   Using Frequency Filters 31
        3.4.2   Using Frequency Filters and Nonlinear Gain 32
    3.5   Dissipative Solitons 33
    3.6   Dispersion-Managed Solitons 35
    3.7   Soliton-Effect Compression 38
    References 39

Chapter 4 Highly Nonlinear Fibers 43

    4.1   Introduction 43
    4.2   Highly Nonlinear Silica Fibers 44
    4.3   Tapered Fibers 44
    4.4   Microstructured Fibers 47

	4.5	Non-silica Fibers ... 50
	4.6	Soliton Fission and Dispersive Waves ... 51
	4.7	Four-Wave Mixing ... 53
	References ... 54	

Chapter 5 Supercontinuum Generation ... 59

- 5.1 Introduction ... 59
- 5.2 Pumping with Picosecond Pulses ... 59
- 5.3 Pumping with a Continuous Wave ... 61
- 5.4 Pumping with Femtosecond Pulses ... 62
- 5.5 Supercontinuum Coherence ... 67
- 5.6 The Supercontinuum as a Source for WDM Systems ... 68
- References ... 69

Chapter 6 Optical Pulse Amplification ... 73

- 6.1 Introduction ... 73
- 6.2 Fiber Raman Amplifiers ... 73
- 6.3 Fiber Brillouin Amplifiers ... 78
- 6.4 Fiber Parametric Amplifiers ... 82
- References ... 86

Chapter 7 All-Optical Switching ... 91

- 7.1 Introduction ... 91
- 7.2 SPM-Induced Optical Switching ... 91
- 7.3 XPM-Induced Optical Switching ... 95
- 7.4 Optical Switching Using FWM ... 98
- References ... 101

Chapter 8 Wavelength Conversion ... 103

- 8.1 Introduction ... 103
- 8.2 FWM-Based Wavelength Converters ... 103
- 8.3 XPM-Based Wavelength Converters ... 109
- References ... 113

Chapter 9 Optical Regeneration ... 117

- 9.1 Introduction ... 117
- 9.2 2R Regenerators ... 117
  - 9.2.1 SPM-Based Regenerators ... 117
  - 9.2.2 SPM-Based Optical Pulse Train Generation ... 120

		9.2.3	Femtosecond Pulse Generation	121
		9.2.4	FWM-Based 2R Regenerators	121
	9.3	3R Regenerators		122
	9.4	All-Optical Regeneration of Phase-Encoded Signals		123
	References			124

Index 129

Author

Mário F. S. Ferreira received his PhD degree in 1992 in physics from the University of Aveiro, Portugal, where he is now a professor in the Physics Department. Between 1990 and 1991, he was at the University of Essex, UK, performing experimental work on external cavity semiconductor lasers and nonlinear optical fiber amplifiers. His research interests have been concerned with the modeling and characterization of multisection semiconductor lasers for coherent systems, quantum well lasers, optical fiber amplifiers and lasers, soliton propagation, nanophotonics, optical sensors, polarization, and nonlinear effects in optical fibers. He has written more than 400 scientific journal and conference publications and several books.

1 Introduction

Ultrafast optical networks are growing at incredible rates, driven by both an increase in the number of connections and the demand for higher bandwidth applications, mainly video content. Dense wavelength-division multiplexing (WDM) technology has been playing one of the most important roles to support this evolution by providing a rapid and timely capacity increase. Today, commercial long-haul C+L-band lightwave systems carry up to 192 channels at up to 250 Gb/s on a 50-GHz grid, corresponding to an aggregate long-haul capacity of ~48 Tb/s and for short-reach applications up to 400 Gb/s, for a total capacity of up to 76 Tb/s [1,2]. Research records have achieved a net per-carrier interface rate in excess of 1 Tb/s [3,4], and aggregate WDM capacities in single-mode fiber of up to 115 Tb/s have been reported [5,6].

One of the key issues in such ultrafast optical networks is how we can achieve various kinds of signal processing functions, such as amplification, reshaping and regeneration, clock extraction and retiming, wavelength conversion, and switching. One approach to realize these functions is the electronic signal processing relying on optoelectronic/electro-optic conversion. The most flexible functions are available in the electronic domain; however, this approach presents bandwidth limitations, prohibitive cost, and manufacturing complexity of high-speed optoelectronics, which become more important as data modulation rates increase.

The all-optical nonlinear approach conveniently avoids the need for optical to electronic conversions, and therefore, circumvents the above limitations. In this case, ultrafast nonlinear optical effects are applied to signal processing, and the response speed can go beyond 1 THz. Moreover, since optical techniques do not need to operate on every individual bit, as electronic transistors do, a single photonic element has the ability to transparently process a data channel regardless of its data rate and modulation format, allowing also for simultaneous operation on multiple data channels [7–9].

Various optical materials and devices have been utilized for optical signal processing, including optical fibers [10–12], semiconductor optical amplifiers (SOAs) [13–15], silicon waveguides [16,17], highly nonlinear waveguides [8,18,19], periodically poled lithium niobate (PPLN) waveguides [20,21], and photonic crystals [22]. Regarding the two first options, while the speed of SOAs is ultimately limited by carrier recovery effects, optical fiber exploiting the third-order $\chi^{(3)}$ optical nonlinearity is nearly instantaneous. Moreover, fiber-based solutions have the advantage of being directly integrated with existing fiber networks and utilizing cheap fiber components for their implementations.

Fiber nonlinearity is responsible for a wide range of phenomena, which can be used to construct a great variety of all-optical signal processing devices. The basic nonlinear phenomena occurring in optical fibers can be classified into two general categories [23]. The first category arises from modulation of the refractive index by intensity changes in the signal (Kerr effect), giving rise to effects such as self-phase modulation (SPM), cross-phase modulation (XPM), and four-wave mixing (FWM). The second category of nonlinearities corresponds to stimulated scattering processes, such as stimulated Brillouin scattering and stimulated Raman scattering. The main characteristics of these effects will be reviewed in Chapter 2 of this book.

Considering the device dimensions, while SOA and PPLN have typical lengths on the order of centimeters, fiber devices based on common dispersion-shifted fiber usually require a length of several kilometers due to its relatively small nonlinear coefficient γ of ~1–2 W^{-1}/km at approximately 1550 nm [10,24–26]. Such a long fiber produces a bulky device, which presents serious limitations concerning its sensitivity to environmental perturbations.

To shorten the length of interaction, highly nonlinear silica fibers with a larger nonlinear coefficient γ of ~10–20 W^{-1}/km were developed through a combination of a smaller effective mode area and enhanced Kerr coefficient n_2 by higher GeO_2 doping of the core. With such fibers, shorter lengths of hundreds of meters are usually used for the demonstration of different nonlinear effects [27–29].

The fiber nonlinearity can be further enhanced by tailoring appropriately its structure. Different types of silica-based microstructured optical fibers (MOFs) have been designed to address this purpose [30]. Fibers with a small core dimension and a cladding with a large air-fill fraction allow for extremely tight mode confinement, that is, a small effective mode area, and hence, a higher value of γ. Using this approach, it has been possible to fabricate pure-silica MOFs with a nonlinear coefficient $\gamma \approx 100$ W^{-1}/km [31]. A 64-m-long silica MOF has been employed for wavelength conversion of communication signals by FWM [32], offering a reduction in fiber length by one order of magnitude.

Significantly higher values of γ can be achieved by combining tight mode confinement with the use of glasses with a greater intrinsic material nonlinearity coefficient than that of silica, namely, including sulfur hexafluoride, lead silicate, tellurite, bismuth oxide, and chalcogenide glass. In particular, chalcogenide glasses provide a very wide transparency window up to 10 µm or 15 µm in the mid-infrared, depending on the glass composition, and a very high third-order nonlinearity, up to 1000 times the one of silica. Using these materials, the required fiber length for several nonlinear processing applications can be impressively reduced to the order of centimeters, representing a further improvement by several orders of magnitude. Impressive results were achieved with bismuth oxide [18,33–35], tellurite [36], and chalcogenide [37,38] glass fibers.

The nonlinear parameter γ of optical fibers made with such high nonlinear materials can be further enhanced by tailoring appropriately their structure [39] or tapering their waist diameter [40,41]. Extreme tapering of a 165-µm diameter As2Se3 fiber produced a dramatic increase in γ from 1200 to 6800 W^{-1}/km [42].

This book is intended to provide an overview of the main all-optical signal processing functions, which can be performed using nonlinear effects in optical fibers, namely pulse amplification and compression, supercontinuum generation and spectrum slicing, optical switching, wavelength conversion, and optical regeneration.

REFERENCES

1. J. Cho, X. Chen, S. Chandrasekhar, G. Raybon, R. Dar, L. Schmalen, E. Burrows, A. Adamiecki, S. Corteselli, Y. Pan, D. Correa, B. McKay, S. Zsigmond, P. Winzer, and S. Grubb, *J. Lightwave Technol.* **36**, 103 (2018).
2. T. Zami, B. Lavigne, O. B. Pardo, S. Weisser, J. David, M. Le Monnier, and J. Faure, *Proc. Optical Fiber Comm. Conf. (OFC)*, W1B.5 (2018).
3. K. Schuh, F. Buchali, W. Idler, T. A. Eriksson, L. Schmalen, W. Templ, L. Altenhain, U. Dümler, R. Schmid, M. Möller, and K. Engenhardt, *Proc. Optical Fiber Comm. Conf. (OFC)*, Th5B.5 (2017).
4. X. Chen, S. Chandrasekhar, G. Raybon, S. Olsson, J. Cho, A. Adamiecki, and P. Winzer, *Proc. Optical Fiber Comm. Conf. (OFC)*, Th4C.1 (2018).
5. A. Sano, T. Kobayashi, S. Yamanaka, A. Matsuura, H. Kawakami, Y. Miyamoto, K. Ishihara, and H. Masuda, *Proc. Optical Fiber Comm. Conf. (OFC)*, PDP5C.3 (2012).
6. J. Renaudier, A. C. Meseguer, A. Ghazisaeidi, P. Tran, R. R. Muller, R. Brenot, A. Verdier, F. Blache, K. Mekhazni, B. Duval, H. Debregeas, M. Achouche, A. Boutin, F. Morin, L. Letteron, N. Fontaine, Y. Frignac, and G. Charlet, *Proc. European Conf. on Optical Comm. (ECOC)*, Th. PDP.A.3 (2017).
7. P. Martelli, P. Boffi, M. Ferrario, L. Marazzi, P. Parolari, R. Siano, V. Pusino, P. Minzioni, I. Cristiani, C. Langrock, M. M. Fejer, M. Martinelli, and V. Degiorgio, *Opt. Express* **17**, 17758 (2009).
8. M. Galili, J. Xu, H. C. Mulvad, L. K. Oxenløwe, A. T. Clausen, P. Jeppesen, B. Luther-Davies, S. Madden, A. Rode, D.-Y. Choi, M. Pelusi, F. Luan, and B. J. Eggleton, *Opt. Express* **17**, 2182 (2009).
9. T. D. Vo, M. D. Pelusi, J. Schroder, F. Luan, S. J. Madden, D.-Y. Choi, D. A. P. Bulla, B. Luther-Davies, and B. J. Eggleton, *Opt. Express* **18**, 3938 (2010).
10. A. Bogoni, L. Potì, R. Proietti, G. Meloni, F. Ponzini, and P. Ghelfi, *Electron. Lett.* **41**, 435 (2005).
11. M. D. Pelusi, F. Luan, E. Magi, M. R. Lamont, D. J. Moss, B. J. Eggleton, J. S. Sanghera, L. B. Shaw, and I. D. Aggarwal, *Opt. Express* **16**, 11506 (2008).
12. S. Radic, *IEEE J. Sel. Topics Quantum Electron.* **18**, 670 (2012).
13. S. Nakamura, Y. Ueno, and K. Tajima, *IEEE Photon. Technol. Lett.* **13**, 1091 (2001).
14. J. Leuthold, L. Moller, J. Jaques, S. Cabot, L. Zhang, P. Bernasconi, M. Cappuzzo, L. Gomez, E. Laskowski, E. Chen, A. Wong-Foy, and A. Griffin, *IEEE Electron. Lett.* **40**, 554 (2004).
15. E. Tangdiongga, Y. Liu, H. de Waardt, G. D. Khoe, A. M. J. Koonen, H. J. S. Dorren, X. Shu, and I. Bennion, *Opt. Lett.* **32**, 835 (2007).
16. M. A. Foster, A. C. Turner, R. Salem, M. Lipson and A. L. Gaeta, *Opt. Express* **15**, 12949 (2007).
17. J. Leuthold, C. Koos, and W. Freude, *Nature Photon.* **4**, 535 (2010).
18. F. Parmigiani, S. Asimakis, N. Sugimoto, F. Koizumi, P. Petropoulos, and D. J. Richardson, *Opt. Express* **14**, 5038 (2006).
19. M. R. E. Lamont, V. G. Ta'eed, M. A. F. Roelens, D. J. Moss, B. J. Eggleton, D.-Y. Choy, S. Madden, and B. Luther-Davies, *Electron. Lett.* **43**, 945 (2007).
20. Y. Fukuchi, T. Sakamoto, K. Taira, and K. Kikuchi, *Electron. Lett.* **39**, 789 (2003).
21. C. Langrock, S. Kumar, J. E. McGeehan, A. E. Willner, and M. M. Fejer, *J. Lightwave. Technol.* **24**, 2579 (2006).
22. J. C. Knight and D. V. Skryabin, *Opt. Express* **15**, 15365 (2007).
23. M. F. Ferreira, *Nonlinear Effects in Optical Fibers*; John Wiley & Sons, Hoboken, NJ (2011).
24. K. Inoue and H. Toba, *IEEE Photon. Technol. Lett.* **4**, 69 (1992).

25. P. Ohlen, B. E. Olsson, and D. J. Blumenthal, *IEEE Photon. Technol. Lett.* **12**, 522 (2000).
26. B. E. Olsson and D. J. Blumenthal, *IEEE Photon. Technol. Lett.* **13**, 875 (2001).
27. H. Sotobayashi, C. Sawaguchi, Y. Koyamada, and W. Chujo, *Opt. Lett.* **27**, 1555 (2002).
28. J. Li, B. E. Olsson, M. Karlsson, and P. A. Andrekson, *IEEE Photon. Technol. Lett.* **15**, 1770 (2003).
29. J. Hansryd, P. A. Andrekson, M. Westlund, J. Li, and P.-O. Hedekvist, *IEEE J. Sel. Topics Quantum Electron.* **8**, 508 (2002).
30. P. S. J. Russel, *J. Lightwave Technol.* **24**, 4729 (2006).
31. J. H. Lee, Z. Yusoff, W. Belardi, M. Ibsen, T. M. Monro, and D. J. Richardson, *IEEE Photon. Technol. Lett.* **15**, 437 (2003).
32. K. K. Chow, C. Shu, C. Lin, and A. Bjarklev, *IEEE Photon. Technol. Lett.* **17**, 624 (2005).
33. J. H. Lee, T. Tanemura, K. Kikuchi, T. Nagashima, T. Hasegawa, S. Ohara, and N. Sugimoto, *Opt. Lett.* **30**, 1267 (2005).
34. M. Scaffardi, F. Fresi, G. Meloni, A. Bogoni, L. Potı, N. Calabretta, and M. Guglielmucci, *Opt. Commun.* **268**, 38 (2006).
35. M. Fok and C. Shu, *IEEE J. Selected Top. Quantum Electron.* **14**, 587 (2008).
36. J. Leong, P. Petropoulos, J. Price, H. Ebendorff-Heidepriem, S. Asimakis, R. Moore, K. Frampton, V. Finazzi, X. Feng, T. Monro, D. Richardson, *J. Lightwave Technol.* **24**, 183 (2006).
37. V. G. Ta'eed, L. B. Fu, M. Pelusi, M. Rochette, I. C. M. Littler, D. J. Moss, and B. J. Eggleton, *Opt. Express* **14**, 10371 (2006).
38. R. E. Slusher, G. Lenz, J. Hodelin, J. Sanghera, L. B. Shaw, and I. D. Aggarwal, *J. Opt. Soc. Am. B* **21**, 1146 (2004).
39. T. M. Monro and H. Ebendorff-Heidepriem, *Annu. Rev. Mater. Res.* **36**, 467 (2006).
40. T. A. Birks, W. J. Wadsworth, and P. St. J. Russell, *Opt. Lett.* **25**, 1415 (2000).
41. G. Brambilla, F. Koizumi, V. Finazzi, and D. J. Richardson, *Electron. Lett.* **41**, 795 (2005).
42. E. C. Magi, L. B. Fu, H. C. Nguyen, M. R. E. Lamont, D. I. Yeom, and B. J. Eggleton, *Opt. Express* **15**, 10324 (2007).

2 Nonlinear Effects in Optical Fibers

2.1 INTRODUCTION

Conventional glass fibers for optical communications are made of fused silica and show an attenuation as low as 0.148 dB/km [1] with a broad low-loss optical window that covers about 60 THz, ranging from 1260 to 1675 µm [2]. A number of third-order nonlinear processes can occur in optical fibers [3]; these can grow to appreciable magnitudes over the long lengths available in fibers, even though the nonlinear index of the silica glass is very small ($n_2 = 2.7 \times 10^{-20}$ m^2/W) [4].

Fiber nonlinearities fall into two general categories [3]. The first category of nonlinearities arises from modulation of the refractive index of silica by intensity changes in the signal (Kerr effect). This gives rise to nonlinearities such as self-phase modulation (SPM), whereby an optical signal alters its own phase; cross-phase modulation (XPM), where one signal affects the phases of all others optical signals and vice versa; and four-wave mixing (FWM), whereby signals with different frequencies interact to produce mixing sidebands. The second category of nonlinearities corresponds to stimulated scattering processes, such as stimulated Brillouin scattering (SBS) and stimulated Raman scattering (SRS), which are interactions between optical signals and acoustic or molecular vibrations in the fibers.

The nonlinear effects generated in the fibers severely affect the performance of optical communications systems, since they impose limits on the launched power of the signals, channel bit rate, channel spacing, transmission bandwidth, and hence the entire information capacity of such systems [5–7]. However, the same nonlinear effects also offer a variety of novel possibilities for ultrafast all-optical signal processing, such as switching, wavelength conversion, amplification, and regeneration [3].

2.2 THE KERR EFFECT

Nonlinear effects are attributed to the dependence of the susceptibility on the electric field, which becomes important at high field strengths. As a result, the total polarization vector **P** can be written in the frequency domain as a power series expansion in the electric field vector [8]:

$$\mathbf{P}(r,\omega) = \varepsilon_0 \left[\chi^{(1)} \cdot \mathbf{E} + \chi^{(2)} : \mathbf{EE} + \chi^{(3)} \vdots \mathbf{EEE} + ... \right] = \mathbf{P}_L(r,\omega) + \mathbf{P}_{NL}(r,\omega) \qquad (2.1)$$

where $\chi^{(j)}$ ($j = 1, 2, \ldots$) is the jth-order susceptibility. To account for the light polarization effects, $\chi^{(j)}$ is a tensor of rank $j + 1$. The linear susceptibility $\chi^{(1)}$ determines the linear part of the polarization $\mathbf{P}_L$. On the other hand, terms of second and higher order in Eq. (2.1) determine the nonlinear polarization $\mathbf{P}_{NL}$. Since SiO_2 is a symmetric molecule, the second-order susceptibility $\chi^{(2)}$ vanishes for silica glasses. As a consequence, virtually all nonlinear effects in optical fibers are determined by the third-order susceptibility $\chi^{(3)}$. In time domain, the form of the expansion is identical to Eq. (2.1) if the nonlinear response is assumed to be instantaneous.

The presence of $\chi^{(3)}$ implies that the refractive index depends on the field intensity, I, in the form [3]

$$n = \sqrt{1 + \chi^{(1)} + \frac{3}{2} \frac{\chi^{(3)}}{c\varepsilon_0 n_0} I} \approx n_0 + n_2 I \tag{2.2}$$

where $n_0 = \sqrt{1 + \chi^{(1)}}$ is the linear refractive index and $n_2 = 3\chi^{(3)}/(4c\varepsilon_0 n_0^2)$ is the refractive index nonlinear coefficient, also known as the *Kerr coefficient*.

In the case of silica fibers, we have $n_0 \approx 1.46$ and $n_2 = 2.7 \times 10^{-20}$ m^2/W [4]. Considering a single-mode fiber with an effective mode area $A_{eff} = 50$ μm^2 carrying a power $P = 100$ mW, the nonlinear part of the refractive index is $n_2 I = n_2 (P/A_{eff}) \approx 6.4 \times 10^{-11}$. In spite of this very small value, the effects of the nonlinear component of the refractive index become significant due to very long interaction lengths provided by the optical fibers.

2.3 THE NONLINEAR SCHRÖDINGER EQUATION

The electric field associated with a bit stream propagating in a single-mode fiber along the z-direction can be written in the form

$$E = \frac{1}{2}\left(F(r,\phi)U(z,t)e^{i(\beta_0 z - \omega_0 t)} + c.c.\right) \tag{2.3}$$

where $U(z,t)$ describes the complex amplitude of the field envelope at a distance z inside the fiber, $F(r,\phi)$ gives the transverse distribution of the fundamental fiber mode, and β_0 is the mode propagation constant at the carrier frequency. The vector nature of the waves is neglected in this section, assuming that they are linearly polarized in the same direction.

The only quantity that changes during propagation is the amplitude $U(z,t)$. Since each frequency component of the optical field propagates with a slightly different propagation constant, it is useful to work in the spectral domain. The evolution of a specific spectral component $\tilde{U}(z,\omega)$ is given by

$$\tilde{U}(z,\omega) = \tilde{U}(0,\omega)\exp\left[i\beta(\omega)z - i\beta_0 z\right] \tag{2.4}$$

where $\tilde{U}(0,\omega)$ is the Fourier transform of the input signal $U(z=0,t)$ and $\beta(\omega)$ is the propagation constant, which can be written as the sum of a linear part, β_L, and a nonlinear part, β_{NL}:

$$\beta(\omega) = \beta_L(\omega) + \beta_{NL}(\omega_0) \tag{2.5}$$

with

$$\beta_L(\omega) = \frac{\tilde{n}(\omega)\omega}{c}, \quad \beta_{NL}(\omega_0) = n_2 I \frac{\omega_0}{c} \tag{2.6}$$

$\tilde{n}$ being the effective mode index. In Eq. (2.6), β_{NL} has been written as a frequency-independent quantity, considering that the spectrum of optical pulses used as bits in communication systems is usually much narrower than the carrier frequency ω_0.

We can expand $\beta_L(\omega)$ in a Taylor series around the carrier frequency ω_0:

$$\beta_L(\omega) = \beta_0 + \frac{d\beta_L}{d\omega}(\omega - \omega_0) + \frac{1}{2}\frac{d^2\beta_L}{d\omega^2}(\omega - \omega_0)^2 + \ldots \tag{2.7}$$

where $\beta_0 = \beta_L(\omega_0)$,

$$\frac{d\beta_L}{d\omega} \equiv \beta_1 = \frac{1}{c}\left[\tilde{n} + \omega\frac{d\tilde{n}}{d\omega}\right] = \frac{n_g}{c} = \frac{1}{v_g} \tag{2.8}$$

and

$$\frac{d^2\beta_L}{d\omega^2} \equiv \beta_2 = -\frac{dv_g/d\omega}{v_g^2} \tag{2.9}$$

In Eq. (2.8), $n_g = \tilde{n} + \omega(d\tilde{n}/d\omega)$ and $v_g = c/n_g$ represent the group refractive index and the group velocity, respectively. The parameter β_2 characterizes the *group velocity dispersion* (GVD). In practice, the group dispersion is often characterized by a group delay parameter, D, defined by delay of arrival time in ps unit for two wavelength components separated by 1 nm over a distance of 1 km. It is related to β_2 as

$$D = -\frac{2\pi c \beta_2}{\lambda^2}. \tag{2.10}$$

For a typical single-mode fiber, we have $D = 0$ ps/(nm-km) (16 ps/(nm-km)) at $\lambda = 1.31\,\mu\text{m}$ (1.55 μm).

Using the expansion given by Eq. (2.7) and retaining only terms up to second order in $(\omega - \omega_0)$, substituting Eqs. (2.5)–(2.7) in Eq. (2.4), calculating the derivative $\partial \tilde{U}/\partial z$, and converting the resultant equation into the time domain by using the inverse Fourier transform and the relation

$$\Delta\omega \leftrightarrow i\frac{\partial}{\partial t} \qquad (2.11)$$

we obtain the equation:

$$i\left(\frac{\partial}{\partial z}+\frac{1}{v_g}\frac{\partial}{\partial t}\right)U - \frac{1}{2}\beta_2\frac{\partial^2}{\partial t^2}U + \beta_{NL}U = 0 \qquad (2.12)$$

The amplitude U in Eq. (2.12) can be normalized such that $P = |U|^2$. Making such normalization, we have $\beta_{NL} = \gamma |U|^2$, where

$$\gamma = \frac{\omega_0 n_2}{cA_{eff}} \qquad (2.13)$$

is known as the fiber nonlinear parameter.

Considering a moving frame propagating with the group velocity and using the new time variable

$$\tau = t - \frac{z}{v_g} \qquad (2.14)$$

Equation (2.12) can be written in the form:

$$i\frac{\partial U}{\partial z} - \frac{1}{2}\beta_2\frac{\partial^2 U}{\partial \tau^2} + \gamma |U|^2 U = 0 \qquad (2.15)$$

Equation (2.15) is usually called the nonlinear Schrödinger equation (NLSE) due to its similarity with the Schrödinger equation of quantum mechanics. The NLSE describes the propagation of pulses in optical fibers taking into account both the group-velocity dispersion and the fiber nonlinearity.

2.4 SELF-PHASE MODULATION

Assuming that the pulse temporal width is sufficiently large, we can neglect the dispersive term in the NLSE in Eq. (2.15), which reduces to

$$\frac{\partial U}{\partial z} = i\gamma PU \qquad (2.16)$$

where $P = |U|^2$ is the pulse power. Neglecting the fiber loss, this corresponds to the input power of the pulse. Equation (2.16) has the following general solution for the field amplitude at the output of a fiber of length L:

$$U(L,\tau) = U(0,\tau)\exp\left[i\phi_{NL}(L,\tau)\right] \qquad (2.17)$$

where $U(0,\tau)$ is the input pulse envelope and

$$\phi_{NL}(L,\tau) = \gamma P(\tau)L \tag{2.18}$$

is the nonlinearity-induced phase change. The form of this general solution shows clearly that the fiber nonlinearity modifies the phase shift across the pulse, but not the intensity envelope.

Using Eq. (2.17), the electric field transmitted in the fiber can be written in the form

$$E(L,t) = \frac{1}{2}\left[\hat{E}(0,t)\exp\left\{i\left[(\beta_0 + \gamma P)L - \omega_0 t\right]\right\} + c.c\right] \tag{2.19}$$

where ω_0 is the pulse central frequency. The phase of the wave described by Eq. (2.19) is

$$\phi = (\beta_0 + \gamma P)L - \omega_0 t \tag{2.20}$$

According with Eq. (2.20), the phase alteration due to the nonlinearity is proportional to the power P of the wave. If the incident wave is a pulse with a given power temporal profile, the power variation within the pulse leads to its own phase modulation. Hence, this phenomenon is appropriately called *self-phase modulation* (SPM).

The instantaneous frequency within the pulse described by Eq. (2.19) is given by:

$$\omega(t) = -\frac{\partial \phi}{\partial t} = \omega_0 - \gamma L \frac{\partial P}{\partial t} \tag{2.21}$$

According to Eq. (2.21), in the leading edge of the pulse, where $dP/d\tau > 0$, the instantaneous frequency is downshifted from ω_0, whereas in the tailing edge, where $dP/d\tau < 0$, the instantaneous frequency is upshifted from ω_0. The frequency at the center of the pulse remains unchanged from ω_0. The chirping due to nonlinearity leads to increased spectral broadening.

In the presence of dispersion, the spectral broadening due to SPM determines two situations qualitatively different. In the normal dispersion region (wavelength shorter than the zero-dispersion wavelength, λ_{ZD}), the chirping due to dispersion is to downshift the leading edge and to upshift the trailing edge of the pulse, which is a similar effect as that due to SPM. Thus, in this regime, the chirping due to dispersion and SPM act in the same direction. On the other hand, in the anomalous dispersion region, the chirping caused by dispersion is opposite to that due to SPM. Consequently, nonlinearity and dispersion-induced chirpings can partially or even completely cancel each other. When this cancellation is total, the pulse neither broadens in time nor in its spectrum, and it is called a *soliton*. Optical solitons offer the possibility of transmitting optical pulses over extremely large distances without distortion [9–11].

Besides its fundamental role in the formation of optical solitons, SPM has been widely used to realize various optical signal processing functions, such as 2R optical regeneration [12–18], pulse train generation [17], and femtosecond pulse generation in fiber lasers [19–23].

2.5 CROSS-PHASE MODULATION

Waves with different wavelengths propagating in the same fiber can interact with each other since the refractive index that a wave experiences depends on the intensities of all other waves. Hence, a pulse at one wavelength has an influence on the phase of a pulse at another wavelength. This nonlinear phenomenon is known as cross-phase modulation (XPM), and it can limit significantly the performance of wavelength-division multiplexing (WDM) systems [24–32].

Let us consider two channels propagating at the same time in the fiber. The two optical fields can be written in the form

$$E_j(r,t) = \frac{1}{2} F(r,\phi) U_j(z) \exp\left[i\left(\beta_j z - \omega_j t\right)\right] + c.c. \quad (j=1,2) \tag{2.22}$$

where $F(r,\varphi)$ gives the transverse spatial distribution of the single mode supported by the fiber, $U_j(z)$ is the normalized slowly varying amplitude of the wave with frequency ω_j, and β_j is the propagation constant. The dispersive effects are taken into account by expanding the frequency-dependent propagation constant, given by Eq. (2.7). Retaining terms only up to the quadratic term in such expansion, the following coupled differential equations are obtained for the slowly varying amplitudes U_1 and U_2:

$$\frac{\partial U_1}{\partial z} + \frac{1}{v_{g1}} \frac{\partial U_1}{\partial t} + i \frac{\beta_{21}}{2} \frac{\partial^2 U_1}{\partial t^2} = i\gamma_1 \left[\begin{array}{l} \left(|U_1|^2 + 2|U_2|^2\right)U_1 + U_1^2 U_2^* \\ \exp\{i(\Delta\beta z - \Delta\omega t)\} \end{array} \right] \tag{2.23}$$

$$\frac{\partial U_2}{\partial z} + \frac{1}{v_{g2}} \frac{\partial U_2}{\partial t} + i \frac{\beta_{22}}{2} \frac{\partial^2 U_2}{\partial t^2} = i\gamma_2 \left[\begin{array}{l} \left(|U_2|^2 + 2|U_1|^2\right)U_2 + U_2^2 U_1^* \\ \exp\{i(-\Delta\beta z + \Delta\omega t)\} \end{array} \right] \tag{2.24}$$

where v_{gj} is the group velocity; β_{2j} is the GVD coefficient, $\Delta\omega = \omega_1 - \omega_2$, $\Delta\beta = \beta_1 - \beta_2$; and γ_j is the nonlinear parameter at frequency ω_j, defined as in Eq. (2.13):

$$\gamma_j = \frac{n_2 \omega_j}{c A_{eff}} \tag{2.25}$$

The last terms in Eqs. (2.23) and (2.24) result from the phenomenon of FWM and can be neglected if the phase matching conditions for the occurrence of this phenomenon are not verified. Moreover, assuming that the pulse duration is long or the dispersion in the fiber is low, we can also neglect the linear terms in Eqs. (2.23) and (2.24), leading to the following simplified system of equations:

$$\frac{\partial U_1}{\partial z} = i\gamma_1 (P_1 + 2P_2) U_1 \tag{2.26}$$

$$\frac{\partial U_2}{\partial z} = i\gamma_2 (P_2 + 2P_1) U_2 \qquad (2.27)$$

where $P_j = |U_j|^2$ is the power of the wave with frequency $\omega_j, j = 1, 2$. If, for simplicity, the powers are assumed as constant, the solution of Eq. (2.26) is given by

$$U_1(L) = U_1(0) \exp\{i\gamma_1 (P_1 + 2P_2) L\} \qquad (2.28)$$

A similar solution can be obtained from Eq. (2.27) for U_2. From Eq. (2.28) it is apparent that the phase of the signal at frequency ω_1 is modified not only due to its own power—which corresponds to the SPM effect—but also due to the power of the signal at frequency ω_2. This phenomenon is referred to as XPM, and Eq. (2.28) shows that it is twice as effective as SPM.

The nonlinear phase shift of the signal at frequency ω_1 resulting from the combination of SPM and XPM at the output of a fiber with length L is given by

$$\phi_{1NL} = \gamma_1 (P_1 + 2P_2) L \qquad (2.29)$$

The XPM-induced frequency shift $\Delta\omega_{1XPM}$ is given by

$$\Delta\omega_{1XPM} = -2\gamma_1 L \frac{\partial P_2}{\partial t} \qquad (2.30)$$

Equation (2.30) shows that the part of the signal at ω_1 that is affected by the leading edge of the signal at ω_2 will be down-shifted in frequency, whereas the part overlapping with the trailing edge will be up-shifted in frequency. This determines a spectral broadening of the signal at ω_1, that is twice the spectral broadening caused by SPM.

As described in other chapters of this book, XPM has been widely explored for several optical signal processing functions, such as all-optical switching, [17,33–37], wavelength conversion [38–42], and full pulse regeneration [43–45].

2.6 FOUR-WAVE MIXING

FWM is a parametric process in which four waves or photons interact with each other due to the third-order nonlinearity of the material. As a result, when several channels with frequencies $\omega_1, ..., \omega_n$ are transmitted simultaneously over the same fiber, the intensity dependence of the refractive index leads not only to phase shifts within a channel, as discussed in previous sections, but also gives rise to signals at new frequencies.

The efficiency of the FWM process depends on the relative phase among the interacting optical waves. In the quantum mechanical description, FWM occurs when photons from one or more waves are annihilated and new photons are created at different frequencies. In this process, the rules of conservation of energy and momentum must be fulfilled. The conservation of momentum leads to the phase-matching condition.

Considering the case in which two photons at frequencies ω_1 and ω_2 are annihilated with simultaneous creation of two photons at frequencies ω_3 and ω_4, the conservation of energy imposes the condition

$$\omega_1 + \omega_2 = \omega_3 + \omega_4 \tag{2.31}$$

On the other hand, the phase matching is given by the condition $\Delta k = 0$, where

$$\begin{aligned}\Delta k &= \beta_3 + \beta_4 - \beta_1 - \beta_2 \\ &= (n_3\omega_3 + n_4\omega_4 - n_1\omega_1 - n_2\omega_2)/c\end{aligned} \tag{2.32}$$

The FWM efficiency is significantly higher in dispersion-shifted fibers than in standard fibers in the wavelength region approximately 1.5 µm due to the low value of the dispersion in the first case.

In practice, it is relatively easy to satisfy the phase-matching condition in the degenerate case $\omega_1 = \omega_2$. In this situation, a strong pump at $\omega_1 = \omega_2 \equiv \omega_p$ generates a low-frequency side band at ω_3 and a high-frequency side band at ω_4, when we assume $\omega_4 > \omega_3$. In analogy to Raman scattering, these side bands are referred to as the Stokes and anti-Stokes bands, respectively, which are also often called the *signal* and *idler* bands. The frequency shift of the two side bands is given by

$$\Omega_s = \omega_p - \omega_3 = \omega_4 - \omega_p \tag{2.33}$$

Assuming that the pulse durations are sufficiently long, we can neglect the linear term corresponding to the second-order dispersion in Eq. (2.15), which becomes

$$\frac{\partial U}{\partial z} = i\gamma |U|^2 U \tag{2.34}$$

In order to consider the FWM process and considering the nondegenerate case ($\omega_1 \neq \omega_2$), the amplitude U in Eq. (2.34) is assumed to be the result of the superposition of four waves:

$$U = U_1 e^{i(\beta_1 z - \omega_1 \tau)} + U_2 e^{i(\beta_2 z - \omega_2 \tau)} + U_3 e^{i(\beta_3 z - \omega_3 \tau)} + U_4 e^{i(\beta_4 z - \omega_4 \tau)} \tag{2.35}$$

Substituting Eq. (2.35) into Eq. (2.34) and considering that the four frequencies satisfy Eq. (2.31), the following set of coupled equations is obtained or the normalized amplitudes U_j:

$$\frac{\partial U_1}{\partial z} = i\gamma \left[\left(|U_1|^2 + 2\sum_{j\neq 1}|U_j|^2\right)U_1 + 2U_3 U_4 U_2^* e^{i\Delta k z}\right] \tag{2.36}$$

$$\frac{\partial U_2}{\partial z} = i\gamma \left[\left(|U_2|^2 + 2\sum_{j\neq 2}|U_j|^2\right)U_2 + 2U_3 U_4 U_1^* e^{i\Delta k z}\right] \tag{2.37}$$

$$\frac{\partial U_3}{\partial z} = i\gamma \left[\left(|U_3|^2 + 2\sum_{j \neq 3} |U_j|^2 \right) U_3 + 2U_1 U_2 U_4^* e^{-i\Delta k z} \right] \quad (2.38)$$

$$\frac{\partial U_4}{\partial z} = i\gamma \left[\left(|U_4|^2 + 2\sum_{j \neq 4} |U_j|^2 \right) U_4 + 2U_1 U_2 U_3^* e^{-i\Delta k z} \right] \quad (2.39)$$

where Δk is the wave-vector mismatch, given by Eq. (2.32), and γ is an averaged nonlinear parameter, which ignores the small variation due to the slightly different frequencies of the involved waves. In deriving Eqs. (2.36)–(2.39), only nearly phase-matched terms were kept. Fiber loss may be included by adding the term $-(\alpha/2)U_j, j = 1, 2, 3, 4$, to the right-hand side of each equation, respectively. The first term inside the brackets in Eqs. (2.36)–(2.39) describes the effect of SPM, whereas the second term is responsible for XPM. Since these terms can lead to a phase alteration only, the generation of new frequency components is provided by the remaining FWM terms. If only the two pump waves (with frequencies ω_1 and ω_2) are present at the fiber input, the signal and idler waves are generated from the wide band noise, which are always present in communication systems. However, in some cases, the signal wave can already be present at the fiber input. If the phase-matching condition is satisfied, both the signal and the idler waves grow during propagation due to the optical power transferred from the two pump waves.

Let us assume that the pump waves are much more intense than the signal and idler waves and that their phases are not matched. In such case, the power transfer between these waves is very ineffective, and we can assume that the pump waves remain undepleted during the FWM process. The solutions of Eqs. (2.36) and (2.37) are then given by:

$$U_j(z) = \sqrt{P_j} \exp\left[i\gamma \left(P_j + 2P_{3-j} \right) z \right], \quad j = 1, 2 \quad (2.40)$$

where $P_j = |U_j(0)|^2$ are the input pump powers. In this approximation, the pump waves experience only a phase shift due to SPM and XPM. We will consider also the case in which the signal wave has already a finite value at the fiber input. However, the input signal power is assumed to be much less than the input pump powers. In these circumstances, due to the phase mismatch between the involved waves, Eq. (2.38) has the following approximate solution:

$$U_3(z) = \sqrt{P_3} \exp\left[i\gamma 2(P_1 + P_2) z \right] \quad (2.41)$$

Substituting Eqs. (2.40) and (2.41) in Eq. (2.39), we can obtain the following result for the power of the idler:

$$P_4(L) = 4\gamma^2 P_1 P_2 P_3 L^2 \left[\frac{\sin(\kappa L/2)}{\kappa L/2} \right]^2 \quad (2.42)$$

where the parameter

$$\kappa = \Delta k + \gamma(P_1 + P_2) \qquad (2.43)$$

corresponds to the effective phase mismatch and $P_1 + P_2 = P_p$ is the total pump power.

If the phase-matching condition $\kappa = 0$ is fulfilled, Eq. (2.42) shows that the power of the idler increases quadratically with the fiber length. However, in such case it is not reasonable to consider that the amplitudes of the pump and signal waves do remain constant along the fiber. If the phases are not matched, the intensity of the idler wave shows a periodic evolution along the fiber, as described by Eq. (2.42).

Let us consider the degenerate case $\omega_1 = \omega_2 \equiv \omega_p$, where ω_p denotes the pump frequency. In this case, the effective phase mismatch is given by

$$\kappa = \Delta k + 2\gamma P_p \qquad (2.44)$$

where P_p is the pump power and Δk is the linear phase mismatch. The linear phase mismatch given by Eq. (2.32) can be expressed in terms of the frequency shift Ω_s given by Eq. (2.33) if we use an expansion of Δk about the pump frequency ω_p. By retaining up to terms quadratic in Ω_s in this expansion, one obtains

$$\Delta k = \beta_2 \Omega_s^2 \qquad (2.45)$$

where $\beta_2 = d^2\beta/d\omega^2$ is the GVD coefficient at the pump frequency.

For a pump wavelength in the anomalous dispersion region $(\beta_2 < 0)$, the negative value of Δk can be compensated by the fiber nonlinearity. For phase matching, one has

$$\kappa \approx \beta_2 \Omega_s^2 + 2\gamma P_p = 0 \qquad (2.46)$$

From this, we obtain that the frequency shift is given by:

$$\Omega_s = \sqrt{2\gamma P_p / |\beta_2|} \qquad (2.47)$$

When $|\beta_2|$ is small, we must take into account the fourth-order term in the expansion of Δk, which becomes

$$\Delta k \approx \beta_2 \Omega_s^2 + \frac{\beta_4}{12}\Omega_s^4 \qquad (2.48)$$

where $\beta_4 = d^4\beta/d\omega^4$. In this case, the phase-matching condition is given by

$$\kappa \approx \beta_2 \Omega_s^2 + \frac{\beta_4}{12}\Omega_s^4 + 2\gamma P_0 = 0 \qquad (2.49)$$

We observe from Eq. (2.49) that phase matching can be achieved even for a pump wavelength in the normal-GVD regime $(\beta_2 > 0)$ if $\beta_4 < 0$. This condition can be easily realized in tapered and microstructured fibers [46,47].

FWM in optical fibers has been widely explored for various functionalities of optical signal processing, such as all-optical switching [48–52], wavelength conversion [53–57], pulse regeneration [58–61], and optical phase conjugation (OPC) [62]. In particular, OPC offers a great potential to compensate Kerr nonlinearity distortions in high-capacity WDM systems [63,64]. As described in Chapter 6, FWM can also be used to realize fiber-optical parametric amplifiers (FOPAs). A FOPA provides a high gain, a broad bandwidth, and may be tailored to operate at any wavelength [65–69].

2.7 STIMULATED RAMAN SCATTERING

In experiences of light scattering in different media, we can observe the existence of a frequency-shifted field in addition to the incident field. The frequency shift of the scattered light is determined by the vibrational oscillations that occur between constituent atoms within the molecules of the material. A slight excitation of the molecular resonances due to the presence of the input field, referred to as the pump wave, results in spontaneous Raman scattering. As the scattered light increases in intensity, the scattering process eventually becomes stimulated. In this regime, the scattered field interacts coherently with the pump field to further excite the resonances, which enhances significantly the transfer of power between the two optical waves. Raman scattering can occur in all materials, but in silica glass the dominant Raman lines are due to the bending motion of the Si–O–Si bond. The theoretical description of SRS has been presented in several publications [70–73].

Both downshift (Stokes) and up-shifted (anti-Stokes) light can, in principle, be generated through the Raman scattering process. However, in the case of optical fibers, the Stokes radiation is generally much stronger than the anti-Stokes radiation due to two main reasons. On one hand, the anti-Stokes process is phase mismatched, whereas the Stokes process is phase matched, for collinear propagation. On the other hand, the anti-Stokes process involves the interaction of the pump light with previously excited molecular resonances, which is not a condition for the Stokes process. Due to these facts, only the Stokes wave will be considered in the following.

Considering the quantum mechanical picture, in the SRS process one has simultaneously the absorption of a photon from the pump beam at frequency ω_p and the emission of a photon at the Stokes frequency, ω_S. The difference in energy is taken up by a high-energy phonon (molecular vibration) at frequency ω_r. Thus, SRS provides for energy gain at the Stokes frequency at the expense of the pump. This process is considered nonresonant because the upper state is a short-lived virtual state.

The pump wave intensity (I_p) and the Stokes wave intensity (I_S) propagating along an optical fiber satisfy the following equations [3]:

$$\frac{dI_S}{dz} = g_R I_S I_p - \alpha I_S \qquad (2.50)$$

$$\frac{dI_p}{dz} = -\frac{\omega_p}{\omega_S} g_R I_S I_p - \alpha I_p \qquad (2.51)$$

where α takes into account the fiber losses, and g_R is the Raman gain coefficient. Figure 2.1 shows the normalized Raman gain spectra for bulk-fused silica, measured when the pump and signal light were either copolarized (full curve) or orthogonally polarized (dotted curve) [74]. The most significant feature of the Raman gain in silica fibers is that g_R extends over a large frequency range (up to 40 THz). Optical signals whose bandwidths are of this order or less can be amplified using the Raman effect if a pump wave with the right wavelength is available. The Raman shift, corresponding to the location of the main peak in Figure 2.1, is close to 13 THz for silica fibers. The Raman gain depends on the relative state of polarization of the pump and signal fields. The copolarized gain is almost an order of magnitude larger than the orthogonally polarized gain near the peak of the gain curve. The peak value of the Raman gain decreases with increasing the pump wavelength, and it is about 6×10^{-14} m/W in the wavelength region around 1.5 μm.

Oxide glasses used as dopants of pure silica in fiber manufacture exhibit Raman gain coefficients that can be significantly higher than that of silica. The most important dopant in communications fibers is GeO_2, whose Raman gain coefficient is about 8.2 times that for pure SiO_2 glass. The Raman gain coefficient increases above that of pure SiO_2 in proportion to the concentration GeO_2 [75]. Besides, doping the core of a silica fiber with GeO_2 also raises the relative index difference, Δ, between the core and cladding, which increases the effective waveguiding and in turn reduces the effective cross-sectional area A_{eff} [76]. As a consequence, the rate of SRS is increased.

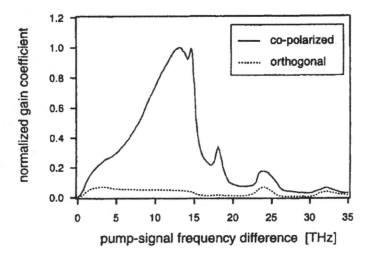

FIGURE 2.1 Normalized Raman gain coefficient for copolarized (full curve) and orthogonally polarized (dotted curve) pump and signal beams. (From J. Bromage, *J. Lightwave Technol.* 22, 79, 2004. © 2004 IEEE. With permission [74].)

When the input Stokes wave intensity is weak, such that $I_{S0} \ll I_{p0}$, the evolution of the Stokes wave intensity is given approximately by:

$$I_S(z) \approx I_{S0} \exp\left\{\frac{g_R I_{p0}\left[1-\exp(-\alpha z)\right]}{\alpha} - \alpha z\right\} \qquad (2.52)$$

In the absence of an input signal I_{S0}, the Stokes wave arises from spontaneous Raman scattering along the fiber. It was shown that this process is equivalent to injecting one fictitious photon per mode at the input end of the fiber [77]. The threshold for SRS is defined as the input pump power at which the output powers for the pump and Stokes wave become equal. It is given approximately by [77]:

$$P_{p0}^{th} = 16 \frac{A_{eff}}{g_R L_{eff}} \qquad (2.53)$$

where

$$L_{eff} = \frac{1-\exp(-\alpha L)}{\alpha} \qquad (2.54)$$

is the effective fiber length. Equation (2.53) was derived assuming that the polarization of the pump and Stokes waves are maintained along the fiber. However, in the case of standard single-mode fibers, due to the arbitrary distribution of the polarization states of both waves, the Raman threshold is increased by a factor of 2. For example, the threshold for the SRS is $P_{p0}^{th} \approx 600$ mW at $\lambda_p = 1.55$ μm in long polarization-maintaining fibers, such that $L_{eff} \approx 22$ km, considering an effective core area of $A_{eff} = 50$ μm^2. However, in standard single-mode fibers with similar characteristics, the threshold would be $P_{p0}^{th} \approx 1.2$ W.

Because SRS has a relatively high threshold, it is not of concern for single-channel systems. However, in WDM systems, SRS can cause crosstalk between channels signals whose wavelength separation falls within the Raman gain curve. Specifically, the long-wavelength signals are amplified by the short-wavelength signals, leading to power penalties for the latter signals. The shortest-wavelength signal is depleted most because it acts as a pump for all other channels. The Raman-induced power transfer between two channels depends on the bit pattern, which leads to power fluctuations and determines additional receiver noise. The magnitude of these deleterious effects depends on several parameters, like the number of channels, their frequency spacing, and the power in each of them.

If the first-order Stokes wave is sufficiently strong, it can act as a pump source itself, resulting in the generation of a higher-order Stokes waves. A sixth-order cascaded SRS in optical fibers was observed for the first time in a 1978 experiment [78]. Since then, many researchers have devoted their attention to the study of cascaded SRS in optical fibers in the visible and near-infrared fields [79–82].

Compared with silica, chalcogenide glasses have wider transmission windows (from the visible up to the infrared region of 12 or 15 μm depending on the composition), and higher Raman gain coefficients ($4.3-5.7\times10^{-12}$ m/W for As_2S_3 at 1.5 μm, and $2-5\times10^{-11}$ m/W for As_2Se_3 at 1.5 μm) [83, 84]. As a result, optical fibers fabricated based on chalcogenide glasses are more promising for generating cascaded SRS in the mid-infrared region [85–87]. This may be particularly useful for some applications, including light detection and ranging (LIDAR), gas sensing, and optical communications.

SRS in optical fibers plays a significant role the supercontinuum generation [3]. Moreover, as will be seen in Chapter 6, the same effect can be used with advantage to implement different types of both lumped and distributed wideband optical amplifiers [88–94].

2.8 STIMULATED BRILLOUIN SCATTERING

Brillouin scattering is a phenomenon named after the French physicist Leon Brillouin, who investigated the scattering of light at acoustic waves during the 1920s. For low incident intensities, the scattered part of the field remains very weak. However, the process becomes stimulated and strong scattered fields are generated at high input intensities, which are readily available with lasers. SBS in optical fibers was observed for the first time in 1972 [95], using a pulsed narrowband Xenon laser operating at 535.3 nm.

The SBS process can be described as a classical three-wave interaction involving the incident (pump) wave of frequency ω_p, the Stokes wave of frequency ω_s, and an acoustic wave of frequency ω_a. The pump creates a pressure wave in the medium owing to electrostriction, which in turn causes a periodic modulation of the refractive index. This process is illustrated in Figure 2.2. Physically, each pump photon in the SBS process gives up its energy to simultaneously create a Stokes photon and an acoustic phonon.

The three waves involved in the SBS process must conserve both the energy and the momentum. The energy conservation requires that $\omega_p - \omega_s = 2\pi f_a$, where f_a is the linear frequency of the acoustic wave, which is about 11.1 GHz in standard fibers. The momentum conservation requires that the wave vectors of the three waves satisfy $\mathbf{k}_a = \mathbf{k}_p - \mathbf{k}_s$. In a single-mode fiber, optical waves can propagate only along the direction of the fiber axis. Since the acoustic wave velocity $v_a \approx 5.96$ km/s is by far smaller than the light velocity, $|\mathbf{k}_a| = 2\pi f_a / v_a > |\mathbf{k}_p| \approx |\mathbf{k}_s|$. In this case, the momentum conservation has the important consequence that the Brillouin effect occurs only if the Stokes and the pump waves propagate in opposite directions.

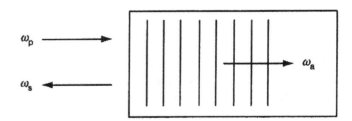

FIGURE 2.2 Schematic illustration of the stimulated Brillouin scattering process.

Nonlinear Effects in Optical Fibers

The response of the material to the interference of the pump and Stokes fields tends to increase the amplitude of the acoustic wave. Therefore, the beating of the pump wave with the acoustic wave tends to reinforce the Stokes wave, whereas the beating of the pump wave and the Stokes waves tends to reinforce the acoustic wave. This explains the appearance of the SBS process.

The pump wave intensity (I_p) and the Stokes wave intensity (I_S) propagating along an optical fiber satisfy the following equations [3]:

$$\frac{dI_S}{dz} = -g_B I_p I_S + \alpha I_S \tag{2.55}$$

$$\frac{dI_p}{dz} = -g_B I_S I_p - \alpha I_p \tag{2.56}$$

where g_B is the Brillouin gain coefficient, which for typical fibers is estimated to be about 2.5×10^{-11} m/W. This value is between two and three orders of magnitude larger than the Raman gain coefficient at $\lambda_p = 1.55$ μm.

The spectrum of the Brillouin gain is Lorentzian with a FWHM Δv_B determined by the acoustic attenuation. For 1.5 μm light, the width is about 17 MHz. However, the gain bandwidth Δv_B can vary from fiber to fiber because of the guided nature of light and the presence of dopants in the fiber core. In particular, the inhomogeneities in the core section along the length determine an increase of the amplifier bandwidth, which can exceed 100 MHz, although typical values are 50–60 MHz for $\lambda_p \approx 1.55$ μm.

In the absence of an input signal, the Stokes wave builds up from spontaneous scattering. In a treatment analogous to that of Raman scattering, the noise power provided by spontaneous Brillouin scattering is equivalent to injecting a fictitious photon per mode at a distance where the gain is equal to the fiber loss [77]. The threshold pump power is defined as the input pump power that is equal to the output power of the Stokes wave, which gives the result [77]:

$$P_{p0}^{th} = 21 \frac{A_{eff}}{g_B L_{eff}} \tag{2.57}$$

Considering an effective area $A_{eff} = 50$ μm^2, an attenuation constant of $\alpha = 0.2$ dB/km, and a gain coefficient $g_B = 2.5 \times 10^{-11}$ m/W, we obtain a threshold of $P_{p0}^{th} \approx 2$ mW. This value is about three orders of magnitude smaller than the threshold required for Raman scattering, which makes SBS the dominant nonlinear effect in some circumstances.

The value reported above for the Brillouin gain coefficient is valid only when the spectral width of the pump beam (Δv_p) is much narrower that the Brillouin linewidth (Δv_B). When this condition is not verified, the Brillouin gain coefficient is reduced and given by

$$\tilde{g}_B = \frac{\Delta v_B}{\Delta v_B + \Delta v_p} g_B \tag{2.58}$$

SBS can be detrimental to such systems in a number of ways: by introducing a severe signal attenuation, by causing multiple frequency shifts, and by introducing inter-channel crosstalk in bidirectional transmission systems. However, Brillouin gain can also find some useful applications, namely for optical amplification [96,97], lasing [98], channel selection in closely spaced wavelength-multiplexed network [99], optical phase conjugation [100], temperature and strain sensing [101,102], all-optical slow-light control [103,104], optical storage [105], etc.

REFERENCES

1. Y. Chigusa, Y. Yamamoto, T. Yokokawa, T. Sasaki, T. Taru, M. Hirano, M. Kakui, M. Onishi, and E. Sasaoka, *J. Lightwave Technol.* **23**, 3541 (2005).
2. Optical System Design and Engineering Consideration, ITU-T Recommendation Series G Supplement 39, (2006).
3. M. F. Ferreira, *Nonlinear Effects in Optical Fibers*; John Wiley & Sons, Hoboken, NJ (2011).
4. E. Voguel, M. Weber, and D. Krol, *Phys. Chem. Glasses* **32**, 231 (1991).
5. A. Chraplyvy, *J. Lightwave Technol.* **8**, 1548 (1990).
6. P. Mitra and J. Stark, *Nature* **411**, 1027 (2001).
7. M. Wu and W. Way, *J. Lightwave Technol.* **22**, 1483 (2004).
8. P. N. Butcher and D. N. Cotter, *The Elements of Nonlinear Optics*; Cambridge University Press, Cambridge, UK (1990).
9. A. Hasegawa and Y. Kodama, *Solitons in Optical Communications*; Oxford University Press, Oxford, UK (1995).
10. A. Hasegawa (Ed.), *New Trends in Optical Soliton Transmission Systems*; Kluwer, Dordrecht, the Netherlands (1998).
11. L. Mollenauer and J. Gordon, *Solitons in Optical Fibers – Fundamentals and Applications*; Elsevier Academic Press, San Diego, CA (2006).
12. P. V. Mamyshev, *Eur. Conf. Opt. Commun. (ECOC98)*, Madrid, Spain, pp. 475–477 (1998).
13. M. Matsumoto, *J. Lightw. Technol*, **23**, 1472 (2004).
14. P. Johannisson and M. Karlsson, *IEEE Photon. Technol. Lett.* **17**, 2667 (2005).
15. A. G. Striegler and B. Schmauss, *J. Lightw. Technol.* **24**, 2835 (2006).
16. M. Rochette, F. Libin, V. Ta'eed, D. J. Moss, and B. J. Eggleton, *IEEE J. Select. Topics Quantum Electron.* **12**, 736 (2006).
17. K. Igarashi and K. Kikuchi, *IEEE J. Sel. Topics Quantum Electron.* **14**, 551 (2008).
18. M. P. Fok and C. Shu, *IEEE J. Sel. Topics Quantum Electron.* **14**, 587 (2008).
19. Z. Liu, Z. M. Ziegler, L. G. Wright, and F. W. Wise, *Optica* **4**, 649 (2017).
20. Z. Liu, C. Li, Z. Zhang, F. X. Kärtner, and G. Chang, *Opt. Express* **24**, 15328 (2016).
21. W. Liu, S.-H. Chia, H.-Y. Chung, R. Greinert, F. X. Kärtner, and G. Chang, *Opt. Express* **25**, 6822 (2017).
22. H.-Y. Chung, W. Liu, Q. Cao, F. X. Kärtner, and G. Chang, *Opt. Express* **25**, 15760 (2017).
23. H. Chung, W. Liu, Q. Cao, L. Song, F. X. Kärtner, and G. Chang, *Opt. Express* **26**, 3684 (2018).
24. A. R. Chraplyvy and J. Stone, *Electron. Lett.* **20**, 996 (1984).
25. J. Wang and K. Petermann, *J. Lightwave Technol.* **10**, 96 (1992).
26. D. Marcuse, A. R. Chraplyvy, and R. W. Tkach, *J. Lightwave Technol.* **12**, 885 (1994).
27. T. K. Chiang, N. Kagi, M. E. Marhic, and L. G. Kazovsky, *J. Lightwave Technol.* **14**, 249 (1996).

28. G. Belloti, M. Varani, C. Francia, and A. Bononi, *IEEE Photon Technol. Lett.* **10**, 1745 (1998).
29. R. Hui, K. R. Demarest, and C. T. Allen, *J. Lightwave Technol.* **17**, 1018 (1999).
30. L. E. Nelson, R. M. Jopson, A. H. Gnauck, and A. R. Chraplyvy, *IEEE Photon. Technol. Lett.* **11**, 907 (1999).
31. S. Betti and M. Giaconi, *IEEE Photon. Technol. Lett.* **13**, 1304 (2001).
32. H. J. Thiele, R. I. Killey, and P. Bayvel, *Opt. Fiber Technol.* **8**, 71 (2002).
33. K. J. Blow, N. J. Doran, B. K: Nayar, and B. P. Nelson, *Opt. Lett.* **15**, 248 (1990).
34. M. Jino and T. Matsumoto, *Electron. Lett.* **27**, 75 (1991).
35. J. E. Sharping, M. Fiorentino, P. Kumar, and R. S. Windeler, *IEEE Photon. Technol. Lett.* **14**, 77 (2002).
36. J. H. Lee, T. Tanemura, T. Nagashima, T. Hasegawa, S. Ohara, N. Sugimoto, and K. Kikuchi, *Opt. Lett.* **30**, 1267 (2005).
37. R. Salem, A. S. Lenihan, G. M. Carter, and T. E. Murphy, *IEEE Photon. Technol. Lett.* **18**, 2254 (2006).
38. C. McKinstrie, S. Radic, and M. Raymer, *Opt. Express* **12**, 5037 (2004).
39. L. B. Fu, M. Rochette, V. G. Ta'eed, et al. *Opt. Express* **13**, 7637 (2005).
40. M. Galili, L. K. Oxenlowe, H. C. H. Hansen, A. T. Clausen, and P. Jeppesen, *IEEE J. Sel. Top. Quantum Electron.* **14**, 573 (2008).
41. M. Fernández-Ruiz, L. Lei, M. Rochette, and J. Azaña, *Opt. Express*, **23**, 22847 (2015).
42. S. Watanabe, T. Kato, T. Tanimura, C. Schmidt-Langhorst, R. Elschner, I. Sackey, C. Schubert, and T. Hoshida, *Opt. Express*, **27**, 16767 (2019).
43. M. Rochette, J. L. Blows, and B. J. Eggleton, *Opt. Express* **14**, 6414 (2006).
44. J. Suzuki, T. Tanemura, K. Taira, et al. *IEEE Photon. Technol. Lett.* **17**, 423 (2005).
45. M. Daikoku, N. Yoshikane, T. Otani, and H. Tanaka, *J. Light. Technol.* **24**, 1142 (2006).
46. W. J. Wadsworth, N. Joly, J. C. Knight, T. A. Birks, F. Biancalana, and P. St. J. Russell, *Opt. Express* **12**, 299 (2004).
47. G. K. Wong, A. Y. Chen, S. G. Murdoch, R. Leonhardt, J. D. Harvey, N. Y. Joly, J. C. Knight, W. J. Wadsworth, and P. St. J. Russel, *J. Opt. Soc. Am. B* **22**, 2505 (2005).
48. T. Morioka, H. Takara, S. Kawanishi, K. Kitoh, and M. Saruwatari, *Electron. Lett.* **32**, 833 (1996).
49. J. Hansryd and P. A. Andrekson, *IEEE Photon. Technol. Lett.* **13**, 732 (2001).
50. T. Sakamoto, K. Seo, K. Taira, N. S. Moon, and K. Kikuchi, *IEEE Photon. Technol. Lett.* **16**, 563 (2004).
51. R. Tang, J. Lasri, P. S. Devgan, V. Grigoryan, P. Kumar, and M. Vasilyev, *Opt. Express* **13**, 10483 (2005).
52. J. P. Cetina, A. Kumpera, M. Karlsson, and P. A. Andrekson, *Opt. Express* **23**, 33426 (2015).
53. J. Hansryd, P. A. Andrekson, M. Westlund, J. Li, and P. O. Hedekvist, *IEEE J. Sel. Top. Quantum Electron.* **8**, 506 (2002).
54. J. M. Chavez Boggio, J. R. Windmiller, M. Knutzen, R. Jiang, C. Brès, N. Alic, B. Stossel, K. Rottwitt, and S. Radic, *Opt. Express* **16**, 5435 (2008).
55. M. Hirano, T. Nakanishi, T. Okuno, and M. Onishi, *IEEE J. Sel. Top. Quantum Electron.* **15**, 103 (2009).
56. D. Nodop, C. Jauregui, D. Schimpf, J. Limpert, and A. Tünnermann, *Opt. Lett.* **34**, 3499 (2009).
57. J. F. Víctor, F. Rancaño, P. Parmigiani, P. Petropoulos, and D. J. Richardson, *J. Lightwave Technol.* **32**, 3027 (2014).
58. G. Cappellini and S. Trillo, *J. Opt. Soc. Am. B* **8**, 824 (1991).
59. Y. Su, L. Wang, A. Agrawal, and P. Kumar, *Electron. Lett.* **36**, 1103 (2000).
60. M. Matsumoto, *IEEE Photon. Technol. Lett.* **17**, 1055 (2005).
61. M. Matsumoto, *IEEE J. Select. Topics Quantum Electron.* **18**, 738 (2012).

62. B. P. P. Kuo, E. Myslivets, A. O. J. Wiberg, S. Zlatanovic, C. S. Bres, S. Moro, F. Gholami, A. Peric, N. Alic, and S. Radic, *J. Lightwave Technol.* **29**, 516 (2011).
63. K. Solis-Trapala, M. Pelusi, H. N. Tan, T. Inoue, and S. Namiki, *J. Lightwave Technol.* **34**, 431 (2016).
64. I. Sackey, F. Da Ros, J. Karl Fischer, T. Richter, M. Jazayerifar, C. Peucheret, K. Petermann, and C. Schubert, *J. Lightwave Technol.* **33**, 1821 (2015).
65. M. Karlsson, *J. Opt. Soc. Am. B* **15**, 2269 (1998).
66. J. Hansryd and P. A. Andrekson, *IEEE Photon. Technol. Lett.* **13**, 194 (2001).
67. C. J. McKinstrie, S. Radic, and A. R. Chraplyvy, *IEEE J. Sel. Top. Quantum Electron.* **8**, 538 (2002).
68. S. Radic and C. J. McKinstrie, *Opt. Fib. Technol.* **9**, 7 (2003).
69. G. Kalogerakis, M. E. Marhic, K. K. Wong, and L. G. Kazovsky, *J. Lightwave Technol.* **23**, 2945 (2005).
70. R. W. Hellwarth, *Phys. Rev.* **130**, 1850 (1963).
71. Y. R. Chen and N. Bloembergen, *Am. J. Phys.* **35**, 989 (1967).
72. A. Penzkofer, A. Laubereau, and W. Kaiser, *Prog. Quantum Electron.* **6**, 55 (1982).
73. R. H. Stolen, J. P. Gordon, W. J. Tomlinson, and H. A. Haus, *J. Opt. Soc. Am. B* **6**, 1159 (1989).
74. J. Bromage, *J. Lightwave Technol.* **22**, 79 (2004).
75. T. Nakashima, S. Seikai, and M. Nakazawa, *Opt. Lett.* **10**, 420 (1985).
76. N. Shibata, M. Horigudhi, and T. Edahiro, *J. Non. Cryst. Solids* **45**, 115 (1981).
77. R. G. Smith, *Appl. Opt.* **11**, 2489 (1972).
78. L. Cohen and C. Lin, *IEEE J. Quantum Electron.* **14**, 855 (1978).
79. H. Pourbeyram, G. P. Agrawal, and A. Mafi, *Appl. Phys. Lett.* **102**, 201107 (2013).
80. K. Yin, B. Zhang, J. Yao, L. Yang, L. Li, and J. Hou, *IEEE Photonics Technol. Lett.* **28**, 11107 (2016).
81. L. Zhang, H. Jiang, X. Yang, W. Pan, and Y. Feng, *Opt. Lett.* **41**, 215 (2016).
82. S. A. Babin, E. A. Zlobina, S. I. Kablukov, and E. V. Podivilov, *Sci. Rep.* **6**, 22625 (2016).
83. T. M. Monro, *Opt. Lett.* **36**, 2351 (2011).
84. A. Tuniz, G. Brawley, D. J. Moss, and B. J. Eggleton, *Opt. Express* **16**, 18524 (2008).
85. J. Troles, Q. Coulombier, G. Canat, M. Duhant, W. Renard, P. Toupin, L. Calvez, G. Renversez, F. Smektala, M. El Amraoui, J. L. Adam, T. Chartier, D. Mechin, and L. Brilland, *Opt. Express* **18**, 26647 (2010).
86. W. Gao, T. Cheng, X. Xue, L. Liu, L. Zhang, M. Liao, T. Suzuki, and Y. Ohishi, *Opt. Express* **24**, 3278 (2016).
87. J. Yao, B. Zhang, K. Yin, L. Yang, J. Hou, and Q. Lu, *Opt. Express* **24**, 14717 (2016).
88. M. Ikeda, *Opt. Commun.* **39**, 148 (1981).
89. A. R. Chraplyvy, J. Stone, and C. A. Burrus, *Opt. Lett.* **8**, 415 (1983).
90. M. Nakazawa, *Appl. Phys. Lett.* **46**, 628 (1985).
91. M. Nakazawa, T. Nakashima, and S. Seikai, *J. Opt. Soc. Am. B* **2**, 215 (1985).
92. M. L. Dakss and P. Melman, *J. Lightwave Technol.* **3**, 806 (1985).
93. N. A. Olsson and J. Hegarty, *J. Lightwave Technol.* **4**, 391 (1986).
94. Y. Aoki, S. Kishida, and K. Washio, *Appl. Opt.* **25**, 1056 (1986).
95. E. P. Ippen and R. H. Stolen, *Appl. Phys. Lett.* **21**, 539 (1972).
96. R. W. Tkach and A. R. Chraplyvy, *Opt. Quantum Electron.* **21**, S105 (1989).
97. M. F. Ferreira, J. F. Rocha, and J. L. Pinto, *Opt. Quantum Electron.* **26**, 35 (1994).
98. D. Y. Stepanov and G. J. Cowle, *IEEE J. Sel. Top. Quantum Electron.* **3**, 1049 (1997).
99. A. Zadok, A. Eyal, and M. Tur, *J. Lightwave Technol.* **25**, 2168 (2007).
100. S. M. Massey and T. H. Russel, *IEEE J. Sel. Top. Quantum Electron.* **15**, 399 (2009).

101. M. Nikles, L. Thevenaz, P. A. Robert, *Opt. Lett.* **21**, 738 (1996).
102. K. Hotate and M. Tanaka, *IEEE Photon. Technol. Lett.* **14**, 179 (2002).
103. K. Y. Song, M. G. Herraez, and L. Thevenaz, *Opt. Express* **13**, 82 (2005).
104. Z. Zhu, A. M. Dawes, D. J. Gauthier, L. Zhang, A. E. Willner, *J. Lightwave Technol.* **25**, 201 (2007).
105. Z. Zhu, D. J. Gauthier, and R. W. Boyd, *Science* **318**, 1748 (2007).

3 Optical Solitons

3.1 INTRODUCTION

The formation of solitons in optical fibers is the result of a balance between the negative (anomalous) group velocity dispersion (GVD) of the glass fiber, which occurs for wavelengths longer than 1.3 μm in a standard fiber, and the Kerr nonlinearity. The existence of fiber solitons was suggested for the first time by Hasegawa and Tappert [1], and it appeared soon as an ideal solution to the problem of pulse spreading caused by fiber dispersion.

After their first experimental observation by Mollenauer et al. [2], Hasegawa [3] made the imaginative proposal that solitons could be used in all-optical transmission systems based on optical amplifiers instead of regenerative repeaters, which were considered standard until 1990. In particular, he suggested using the Raman effect of the transmission fiber itself for optical amplification. The distributed Raman amplification gave its position to erbium-doped fiber amplifiers (EDFAs) during the 1990s.

Meanwhile, Gordon and Haus [4] anticipated that the transmission of a signal made of optical solitons could not be extended an unlimited distance when optical amplification is used. In fact, the amplifiers needed to compensate for the fiber loss generate amplified spontaneous emission (ASE), and this noise is, in part, incorporated by the soliton, whose mean frequency is then shifted. Due to GVD, the arrival time of the soliton becomes then a random variable, whose variance is proportional to the cube of the propagation distance. This is the so-called *Gordon–Haus effect*.

Some research groups suggested the use of frequency filters to extend the limit set by the Gordon–Haus effect [5–7]. However, using this technique, the transmission distance was still limited by the growth of narrow-band noise at the center frequency of the filters. This problem was beautifully solved by Mollenauer and coworkers, who developed the sliding-guiding filter concept [8–10]. Other proposals to achieve a stable soliton propagation in fiber systems made use of amplitude modulators [11] or nonlinear optical amplifiers [12–14].

The nonlinear pulses propagating in the presence of some of the control techniques referred above are the result of the double balance between nonlinearity and dispersion and between gain and loss. Such nonlinear pulses are known as *dissipative solitons*, and their properties are completely determined by the external parameters of the optical system [15–17].

Attempts to solve much of the intrinsic problems of optical soliton transmission by proper control of the fiber group dispersion—a technique usually referred as *dispersion management*—has emerged in the latter half of the 1990s. In particular, Smith et al. [18] showed in 1996 that a nonlinear soliton-like pulse can exist in a fiber having a periodic variation of the dispersion, even if the dispersion is almost zero. The nonlinear pulse that can propagate in such a system is usually called the *dispersion-managed soliton* and presents several remarkable characteristics, namely, an enhanced pulse energy, reduced Gordon–Haus timing jitter, longer collision lengths, and greater robustness to polarization-mode dispersion (PMD) [18–22]. Due to their characteristics, dispersion-managed solitons are the preferred option for use in new ultrahigh-speed multiplexed systems.

3.2 SOLITON SOLUTIONS OF THE NONLINEAR SCHRÖDINGER EQUATION

Considering Eq. (2.15), let us introduce a normalized amplitude Q given by

$$U(z,\tau) = \sqrt{P_0}\,Q(z,\tau) \tag{3.1}$$

where P_0 is the peak power of the incident pulse. $Q(z,\tau)$ is found to satisfy the equation:

$$i\frac{\partial Q}{\partial z} - \frac{1}{2}\beta_2 \frac{\partial^2 Q}{\partial \tau^2} + \gamma P_0 |Q|^2 Q = 0 \tag{3.2}$$

Let us define a *nonlinear distance*, L_{NL}, as

$$L_{NL} = \frac{1}{\gamma P_0} \tag{3.3}$$

L_{NL} is the propagation distance required to produce a nonlinear phase change rotation of one radian at a power P_0. A *dispersion distance*, L_D, can be also defined as

$$L_D = \frac{t_0^2}{|\beta_2|} \tag{3.4}$$

where t_0 is the input pulse width. These two characteristic distances provide the length scales over which nonlinear or dispersive effects become important for pulse evolution.

Using a distance Z normalized by the dispersion distance L_D and a time scale T normalized by t_0, Eq. (3.2) becomes

$$i\frac{\partial Q}{\partial Z} \pm \frac{1}{2}\frac{\partial^2 Q}{\partial T^2} + N^2 |Q|^2 Q = 0 \tag{3.5}$$

where

$$N^2 = \frac{L_D}{L_{NL}} = \frac{\gamma P_0 t_0^2}{|\beta_2|} \tag{3.6}$$

In the second term of Eq. (3.5), the plus signal corresponds to the case of anomalous GVD $(\text{sgn}(\beta_2) = -1)$, whereas the minus signal corresponds to normal GVD $(\text{sgn}(\beta_2) = +1)$. The parameter N can be eliminated from Eq. (3.5) by introducing a new normalized amplitude $q = NQ$, with which Eq. (3.5) takes the standard form of the nonlinear Schrödinger equation (NLSE):

$$i\frac{\partial q}{\partial Z} + \frac{1}{2}\frac{\partial^2 q}{\partial T^2} + |q|^2 q = 0 \qquad (3.7)$$

The case of anomalous GVD $(\beta_2 < 0)$ was assumed in writing Eq. (3.7).

The NLSE given by Eq. (3.7) is a completely integrable equation that can be solved exactly using the inverse scattering method [23,24], as shown by Zakharov and Shabat [25]. In the case $N = 1$, we have the fundamental soliton solution, which can be written in the form:

$$q(T,Z) = \eta \,\text{sech}\left[\eta\left(T + \kappa Z - T_0\right)\right]\exp\left(-i\kappa T + \frac{i}{2}\left(\eta^2 - \kappa^2\right)Z + i\sigma\right) \qquad (3.8)$$

This solution is characterized by four parameters: the amplitude η (also the pulse width), the frequency κ (also the pulse speed), the time position T_0, and the phase σ. The N-soliton solutions are asymptotically given in the form of N-separated solitons:

$$q(Z,T) = \sum_{j=1}^{N} \eta_j \,\text{sech}\left[\eta_j\left(T + \kappa_j Z - T_{0j}\right)\right]\exp\left[-i\kappa_j T + i\frac{1}{2}\left(\eta_j^2 - \kappa_j^2\right)Z - i\sigma_j\right] \qquad (3.9)$$

If the input pulse shape is symmetrical, all the output solitons propagate at exactly the same speed. This type of solution is called the *bound-soliton solution*, and its shape evolves periodically during the propagation due to the phase interference among the constituting solitons. For example, considering an input pulse shape given by

$$q(0,T) = N\text{sech}(T) \qquad (3.10)$$

the period of oscillation of the higher-order solitons is

$$Z_0 = \frac{\pi}{2} \qquad (3.11)$$

According to Eq. (3.6), the peak power of a soliton in an optical fiber is

$$P_0 = \frac{N^2|\beta_2|}{\gamma t_0^2} \qquad (3.12)$$

Considering a temporal width $t_0 = 6$ ps and using typical parameter values $\beta_2 = -1$ ps^2/km and $\gamma = 3$ W^{-1}/km for dispersion-shifted fibers, we obtain

$P_0 \sim 10$ mW for fundamental solitons ($N = 1$) in the C-band ($\lambda \approx 1550$ nm). The required power to launch the Nth-order soliton is N^2 times that for the fundamental one. Moreover, higher-order solitons compress periodically, resulting in soliton chirping and spectral broadening. In contrast, fundamental solitons preserve their shape during propagation. This fact, together with the lower power required for their generation, makes fundamental solitons the preferred option in signal processing and soliton communication systems.

3.3 PERTURBATIONS OF SOLITONS

High-speed fiber-optic communication systems are generally limited both by the fiber nonlinearity and by the group dispersion, that causes the pulse broadening. However, since fundamental solitons are obtained by the balance between the group dispersion and Kerr nonlinearity, their width can be maintained over long distances. In other words, fundamental solitons are free from either the dispersive distortion or from self-phase modulation. Thus, it is quite natural to use solitons as information carrier in fibers since all other formats will face distortion either from dispersion or nonlinearity.

In a realistic communication system, the soliton propagation is affected by many perturbations related to the input pulse shape, chirp and power, amplifiers noise, optical filters, modulators, etc. However, because of its particle-like nature, it turns out that the soliton remains stable under most of these perturbations.

3.3.1 Fiber Losses

Fiber loss is the first perturbation affecting the soliton propagation and one of the main causes of signal degradation in long-distance fiber-optic communication systems. This limitation can be minimized by operating near $\lambda = 1.55$ μm. However, even with fiber losses as low as 0.2 dB/km, the signal power is reduced by 20 dB after transmission of over 100 km of fiber.

Fiber loss can be taken into account theoretically by adding a loss term to the NLSE, which becomes

$$i\frac{\partial q}{\partial Z} + \frac{1}{2}\frac{\partial^2 q}{\partial T^2} + |q|^2 q = -\frac{i}{2}\Gamma q \qquad (3.13)$$

where $\Gamma = \alpha_0 L_D$ is the loss rate per dispersion distance and α_0 is the fiber loss coefficient. Assuming that $\Gamma \ll 1$ and using the soliton perturbation theory [26], we find that the evolution of the soliton amplitude is given by:

$$\eta(Z) = \eta(0)\exp(-\Gamma Z) \equiv \eta_0 \exp(-\Gamma Z) \qquad (3.14)$$

Equation (3.14) indicates that the soliton amplitude decreases along the fiber length at the same rate as the power amplitude. The decrease in the amplitude at a rate twice as fast as a linear pulse is a consequence of the nonlinear property of a soliton.

Since the amplitude and width are inversely related, the soliton width will increase exponentially according to

$$t_p(Z) = t_0 \exp(\Gamma Z) \tag{3.15}$$

In fact, the exponential increase of the soliton width predicted by the perturbation theory occurs only for relatively short propagation distances. For large distances, the soliton width increases linearly with distance, but at a rate slower than the linear pulse [26,27].

In the context of communications systems, the fiber loss problem was solved in a first stage using repeaters periodically installed in the transmission line. A repeater consists of a light detector and light pulse generators. In fact, it was the most expensive unit in the transmission system, and its use represented also the bottleneck to increase the transmission speed.

Repeaterless soliton transmission systems using distributed Raman gain provided by the fiber itself to compensate for the fiber loss was suggested by Hasegawa [3] in 1983. Its use requires periodic injection of the pump power into the transmission fiber.

The concept of all-optical transmission using the distributed Raman amplification gave its position to lumped amplification during the 1990s, through the use of EDFAs. The EDFA can be pumped by laser diodes and presents some remarkable properties [28]. Using lumped EDFAs, the amplification occurs over a very short distance (~10 m), which compensates for the loss occurring over 40–50 km. However, by lightly doping the transmission fiber, a distributed amplification configuration can also be achieved.

In comparison with the lumped-amplification scheme, the distributed-amplification configuration appears as a better approach since it can provide a nearly lossless fiber by compensating losses locally at every point along the fiber link. In the presence of a distributed gain with $G \ll 1$, such as that provided by the Raman amplification or a distributed EDFA, the soliton amplitude given by Eq. (3.14) is simply modified to

$$\eta(Z) = \eta_0 \exp\left(\int_0^Z [2G(Z) - \Gamma] \, dZ\right). \tag{3.16}$$

By designing the gain so that the exponent of Eq. (3.16) vanishes, one can achieve a system in which a soliton propagates without any distortion.

3.3.2 HIGHER-ORDER EFFECTS

The propagation of short solitons ($t_0 < 5$ ps) in distributed fiber amplifiers can be described by a generalized NLSE which, besides the addition of a gain term, must include also the effects of finite gain bandwidth, third-order dispersion (TOD), and intrapulse Raman scattering (IRS). The inclusion of these effects provides a generalization of Eq. (3.13), in the form [29]:

$$i\frac{\partial q}{\partial Z} + \frac{1}{2}\frac{\partial^2 q}{\partial T^2} + |q|^2 q = -\frac{i}{2}\Gamma q + \frac{i}{2}g(Z)L_D\left(q + \tau_2^2 \frac{\partial^2 q}{\partial T^2}\right) + i\delta_3 \frac{\partial^3 q}{\partial T^3} + \tau_R q \frac{\partial |q|^2}{T} \tag{3.17}$$

where τ_2 is related inversely to the gain bandwidth. The TOD and Raman effects are represented, respectively, by the parameters δ_3 and τ_R, given by

$$\delta_3 = \frac{\beta_3}{6|\beta_2|t_0}, \quad \tau_R = \frac{t_R}{t_0} \tag{3.18}$$

where t_R is a constant dependent on the Raman gain slope. The TOD term becomes important only when $|\beta_2|$ is sufficiently small. On the other hand, since $t_R \approx 5$ fs, the Raman term in Eq. (3.17) can generally be treated as a small perturbation. Considering only this term and applying the perturbation theory, it is found that the soliton amplitude remains constant, whereas its frequency varies with distance as

$$\kappa(Z) = -\frac{8}{15}\tau_R \eta^4 Z. \tag{3.19}$$

The IRS effect leads to a continuous downshift of the carrier frequency, an effect known as the *soliton self-frequency shift* (SSFS) [30]. This effect was observed for the first time by Mitschke and Mollenauer in 1986 using 0.5-ps pulses obtained from a passively mode-locked color-center laser [31]. The origin of SSFS can be understood by noting that for ultrashort solitons, the pulse spectrum becomes so broad that the high-frequency components of the pulse can transfer energy through Raman amplification to the low-frequency components of the same pulse. Such an energy transfer appears as a red shift of the soliton spectrum, with shift increasing with distance. Since the amplitude and width are inversely related, the frequency shift scales as t_0^4, indicating that it can become quite large for ultrashort short pulses. In practice, the limited bandwidth of the gain spectrum $(\tau_2 \neq 0)$ can provide in some circumstances the SSFS compensation and adiabatic soliton trapping [29].

3.3.3 Timing Jitter

The amplifier noise normally deteriorates the signal-to-noise ratio (SNR) for a linear signal due to the superposition of noise with the signal. For soliton transmission, this problem is generally not serious because the signal amplitude can be made large with proper choice of the fiber dispersion. However, the velocity modulation induced by the frequency modulation by the amplifier noise results in jitter in arrival times. This is known as the called Gordon–Haus jitter. The variance of such timing jitter at $Z = Z$ is given by

$$\langle T_0^2 \rangle = \frac{(G-1)\eta}{9 N_0 Z_a} Z^3 \tag{3.20}$$

where G is the amplifier gain, η is the soliton amplitude, Z_a is the amplifier spacing, and N_0 is the number of photons per unit energy. As observed from Eq. (3.20), the Gordon–Haus timing jitter increases with the cube of distance and, in practice, it sets a limit on the bit-rate-distance product of a communication system.

Besides the frequency, the ASE noise added by the amplifiers affects also the other three soliton parameters (amplitude, position, and phase), as well as its polarization.

Optical Solitons

In particular, ASE noise-induced amplitude fluctuations originate also a timing jitter, which, in the case of ultrashort solitons, becomes more important than that due to the Gordon–Haus effect. Such amplitude fluctuations are converted into frequency fluctuations by the IRS effect [see Eq. (3.19)], and give origin to a timing jitter, which increases with the fifth power of distance [32].

Another contribution to the timing jitter, which was observed in the earliest long-distance soliton transmission experiments [33], arises from an acoustic interaction among the solitons. As a soliton propagates down the fiber, an acoustic wave is generated through electrostriction. This acoustic wave induces a variation in the refractive index, which affects the speed of other pulses following in the wake of the soliton. In practice, since a bit stream consists of a random string of 1 and 0 bits, changes in the speed of a given soliton depend on the presence or absence of solitons in the preceding bit slots. The timing jitter arises because, in these circumstances, different solitons experience different changes of speed.

3.4 SOLITON TRANSMISSION CONTROL

Although solitons are clearly a nice option as the information carrier in optical fibers because of their robust nature, their interaction and the timing jitter can lead to the loss of information in transmission systems. These problems can be suppressed using some techniques to control the soliton parameters, namely, its amplitude (or width), time position, velocity (or frequency), and phase. In fact, while the original wave equation has infinite-dimensional parameters, control of finite-dimensional parameters is sufficient to control the soliton transmission system. This is an important characteristic of a soliton system.

3.4.1 Using Frequency Filters

The use of narrow-band filters was early suggested to control the Gordon–Haus jitter and other noise effects [5–7]. The basic idea is that any soliton whose central frequency has strayed from the filter peak will be returned to the peak by virtue of the differential loss the filters induce across its spectrum. Suppression of the timing jitter is a consequence of the frequency jitter damping. In practice, etalon guiding filters are used. The multiple peaks present in the response of these filters provide the possibility of their use in wavelength-division multiplexed (WDM) soliton transmission systems.

Using the soliton perturbation theory, we can obtain the following evolution equations for the soliton amplitude η and the frequency κ in the presence of filters of strength β ($\beta > 0$):

$$\frac{\partial \eta}{\partial Z} = 2\delta\eta - 2\beta\eta\left(\frac{1}{3}\eta^2 + \kappa^2\right) \quad (3.21)$$

$$\frac{\partial \kappa}{\partial Z} = -\frac{4}{3}\beta\eta^2\kappa \quad (3.22)$$

where δ is the excess gain necessary to compensate for the filter-induced loss. From Eq. (3.22), it can be seen that the soliton frequency approaches asymptotically to

$\kappa = 0$ (stable fixed point) if $\eta \neq 1$. In such limit, Eq. (3.21) provides a stationary amplitude $\eta = 1$ if the condition $\beta = 3\delta$ is satisfied. In this case, solitons having an initial range of amplitudes and frequencies emerge as solitons with an identical amplitude and frequency imposed by the attractor at $\eta = 1$ and $\kappa = 0$ after repeated amplifications. This process may be interpreted as soliton cooling and can be used to overcome the Gordon–Haus effect. In fact, the use of fixed-frequency guiding filters provides an effective suppression of the soliton timing jitter, whose variance is reduced relatively to the uncontrolled case, given by Eq. (3.20), by the factor [34]:

$$F_r(x) = \frac{3}{2}\frac{1}{x^3}\left[2x - 3 + 4\exp(-x) - \exp(-2x)\right] \tag{3.23}$$

where $x = 4\delta Z$. If $x \gg 1$, $f(x) \sim 3x^{-2}$, and the variance of the timing jitter increases linearly with distance, instead of the cubic dependence shown in the uncontrolled case [6,7,34].

As mentioned above, to compensate for the filter-induced loss, some excess gain must be provided to the soliton. However, this excess gain amplifies also the small-amplitude waves coexistent with soliton. Such amplification results in a background instability, which can affect significantly and even destroy the soliton itself [26,34].

An approach to avoid the background instability was suggested by Mollenauer et al. [8] and consists in using filters whose central frequency is gradually shifted along the transmission line. In this scheme, the linear waves that initially grow near $\kappa = 0$ eventually fall into negative gain region of the filter and are dissipated while the soliton central frequency is shifted, following the central frequency of the filter. This example represents a remarkable property of solitons, since the signal carried by them can be effectively separated from noise, which has the same frequency components as the signal.

Another technique to control the timing jitter due to the Gordon–Haus effect was suggested by Nakazawa et al. [11] and consists of the use of a modulator, which is timed to pass solitons at the peak of their transmissions. Synchronous modulators work by forcing the soliton to move toward its transmission peak, where loss is minimum, and such forcing reduces timing jitter considerably.

3.4.2 Using Frequency Filters and Nonlinear Gain

An alternative approach to avoid the background instability consists of the use of an amplifier having a nonlinear property of gain, or gain and saturable absorption in combination, such as the nonlinear loop mirror or nonlinear polarization rotation in combination with a polarization-dependent loss element [13]. The key property of the nonlinearity in gain is to give an effective gain to the soliton and a suppression (or very small gain) to the noise. This method of nonlinear gain may be particularly useful for transmission of solitons with subpicosecond or femtosecond durations, where the gain bandwidth of amplifiers will not be wide enough for the sliding of the filter frequency to be allowed.

The pulse propagation in optical fibers where linear and nonlinear amplifiers and narrow-band filters are periodically inserted may be described by the following modified NLSE [35–38]:

$$i\frac{\partial q}{\partial Z} + \frac{1}{2}\frac{\partial^2 q}{\partial T^2} + |q|^2 q = i\delta q + i\beta \frac{\partial^2 q}{\partial T^2} + i\varepsilon |q|^2 q + i\mu |q|^4 q + v|q|^4 q \qquad (3.24)$$

where β stands for spectral filtering, δ is the linear gain or loss coefficient, ε accounts for nonlinear gain-absorption processes, μ represents a higher-order correction to the nonlinear gain absorption, and v is a higher-order correction term to the nonlinear refractive index.

Equation (3.24) is known as the complex Ginzburg–Landau equation (CGLE), so-called cubic for $\mu = v = 0$ and quintic for $\mu, v \neq 0$. It describes to a good approximation the soliton behavior both in optical transmission lines and in mode-locked fiber lasers [39–42].

Using the soliton perturbation theory, Eq. (3.22) is again derived for the soliton frequency κ, whereas the soliton amplitude η satisfies the evolution equation:

$$\frac{\partial \eta}{\partial Z} = 2\delta\eta - 2\beta\eta\left(\frac{1}{3}\eta^2 + \kappa^2\right) + \frac{4}{3}\varepsilon\eta^3 + \frac{16}{15}\mu\eta^5 \qquad (3.25)$$

A stable pulse propagating in a stable background can be achieved if the following conditions are verified [14]:

$$\delta < 0,\ \mu < 0,\ \varepsilon > \beta/2,\ 15\delta > 8\mu\eta_s^4 \qquad (3.26)$$

where η_s is the stationary value for the soliton amplitude. The above conditions show that the inclusion of the quintic term in Eq. (3.24) is necessary to guarantee the stability of the whole solution: pulse and background.

3.5 DISSIPATIVE SOLITONS

Besides the propagation of nonlinear pulses in fiber systems with gain and loss, the CGLE given by Eq. (3.24) has been used to describe many non-equilibrium phenomena, such as convection instabilities, binary fluid convection, and phase transitions [43–45]. In particular, the formation of solitons in systems far from equilibrium has emerged in the last few years as an active field of research [15]. These solitons are termed *dissipative solitons*, and they emerge as a result of a double balance: between nonlinearity and dispersion and also between gain and loss [46]. The properties of dissipative solitons are completely determined by the external parameters of the system, and they can exist indefinitely in time, as long as these parameters stay constant. However, they cease to exist when the source of energy is switched off, or if the parameters of the system move outside the range, which provides their existence.

Even if it is a stationary object, a dissipative soliton shows nontrivial energy flows with the environment and between different parts of the pulse. Hence, this kind of soliton is an object that is far from equilibrium and presents some characteristics similar to a living thing. Like a dissipative soliton, an animal is a localized "structure" that has material and energy inputs and outputs and complicated internal dynamics, which exists only for a certain range of parameters (pressure, temperature, humidity, etc.), and dies if there is no supply of energy. These ideas can be applied to various fiber optic devices, such as passively mode-locked fiber lasers [47,48] and high-density optical transmission lines [49].

As discussed in Section 3.4.2, the parameter space where stable solitons exist has certain limitations. We must have $\beta > 0$ in order to stabilize the soliton in the frequency domain. The linear gain coefficient δ must be negative in order to avoid the background instability. The parameter μ must be negative in order to stabilize the soliton against collapse. Concerning the parameter ν, it can be positive or negative.

Stable dissipative solitons can be found numerically from the propagation Eq. (3.24) taking as the initial condition a pulse of somewhat arbitrary profile. In general, if the result of the numerical calculation converges to a stationary solution, it can be considered as a stable one, and the chosen set of parameters can be deemed to belong to the class of those that permit the existence of solitons.

Different types of soliton solutions were obtained in this way, which can be divided in two main classes: localized fixed-shape solutions and localized pulsating solutions [15,26,50–54]. Examples of localized fixed-shape solutions are the plain pulses (PPs) [26], the flat-top pulses [26,50], the composite pulses [50,55], the dissipative soliton resonance pulses [56–61], and the high-amplitude pulses [62,63]. Stable PPs have a sech profile similar to the soliton solutions of the NLSE and are obtained for small values of the parameters in the right-hand side of Eq. (3.24). The other types of pulses are obtained for nonsmall values of those parameters.

The concept of dissipative soliton resonance (DSR) has also been proposed in the area of soliton fiber lasers to achieve high-energy wave-breaking free pulses [56–61]. With the increase of pump power, the energy of a DSR pulse can increase mainly due to the increase of its pulse width, while keeping the amplitude at a constant level. A different kind of high-energy ultrashort pulses correspond to the high-amplitude soliton solutions of the CGLE found by Latas et al. [62,63]. Such soliton solutions occur due to a singularity first predicted both numerically and using the soliton perturbation theory in Refs. [64,65], namely, as the nonlinear gain saturation effect tends to vanish. The increase in energy of these pulses is mainly due to the increase of the pulse amplitude, whereas the pulse width becomes narrower.

Among the localized pulsating solutions, we may refer the plain pulsating and the creeping solitons, as well as the erupting solitons, which belong to the class of chaotic solutions [51,52]. Figure 3.1 shows an example of the evolution of an erupting pulse, considering the following set of parameters: for $\delta = -0.1$, $\beta = 0.125$, $\varepsilon = 1.0$, $\mu = -0.1$, and $\nu = -0.6$.

The evolution of the pulse starts from a stationary localized solution. After a while, its "slopes" become covered with small ripples (small-scale instability), which seem to move downward along the two slopes of the soliton, as observed in Figure 3.1a. When the ripples increase in size, the soliton cracks into pieces, after

Optical Solitons

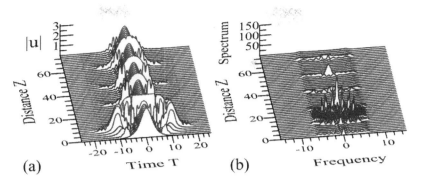

FIGURE 3.1 (a) Amplitude and (b) spectrum of an erupting soliton for the following parameter values: $\delta = -0.1$, $\beta = 0.125$, $\varepsilon = 1.0$, $\mu = -0.1$, and $\nu = -0.6$.

which the pulse evolves in order to restore its shape. In the spectral domain, the pulse exhibits a dual-peak spectrum, which evolves during propagation as illustrated in Figure 3.1b. Some perturbations appear at the central frequency, well separated from each of the two main peaks. These perturbations extend to the entire spectrum just before explosions occur. This process repeats forever, but never exactly in the same way and with equal periods. In fact, the erupting soliton solution belongs to the class of chaotic solutions.

The existence of the erupting solitons has been experimentally confirmed in a passively mode-locked solid-state laser, where the higher-order effects might have some influence [66]. The interaction among such higher-order effects, namely, the third-order dispersion, the intrapulse Raman scattering, and the self-steepening effects, becomes particularly important for stable femtosecond pulse propagation and generation by passively mode-locked lasers [26]. It has been shown that if such higher-order effects act together, the explosions of an erupting soliton can be drastically reduced and even eliminated, providing also a fixed-shape pulse [67–69].

3.6 DISPERSION-MANAGED SOLITONS

Various dispersion profiles (or dispersion maps) have been tried in linear transmission systems to minimize the nonlinear effects. Basically, the idea is to introduce locally relatively large dispersions to make nonlinear effects relatively less important and to compensate for the accumulated dispersion at the end of the line so that the integrated dispersion becomes zero. A method of programming dispersion is called *dispersion management*. Dispersion management was also found to be effective in soliton transmission systems since it can reduce significantly the timing jitter, which is the major cause of bit error in such systems.

Smith et al. [18] proposed the use of a periodic map using both anomalous dispersion and normal dispersion fibers alternately. The nonlinear stationary pulse that propagates in such a fiber has a peak power larger than the soliton that propagates in a fiber with a constant dispersion equal to the average dispersion of the period map. Such a nonlinear stationary pulse is commonly called a dispersion-managed (DM) soliton, and it has been introduced simultaneous by many authors [22].

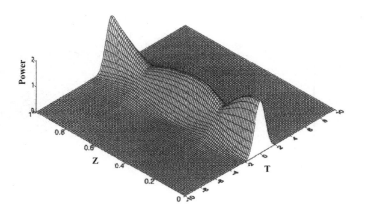

FIGURE 3.2 Evolution of the DM soliton along one dispersion map period.

DM solitons are entities whose shape varies substantially over one period of the dispersion map but which return to the same shape at the end of each period. Figure 3.2 shows the evolution of the DM-soliton during one period of the dispersion map with anomalous- and normal-dispersion fibers of equal length. The DM-pulse alternately spreads and compresses as the sign of the dispersion is switched. The pulse peak power varies rapidly, and the pulse width becomes minimum at the center of each fiber where frequency chirp vanishes.

Propagation in a DM transmission line, which includes optical amplifiers, with gain $G(Z)$ to compensate for the loss with loss rate Γ is described by the equation:

$$i\frac{\partial q}{\partial Z} + \frac{1}{2}d(Z)\frac{\partial^2 q}{\partial T^2} + |q|^2 q = i\big[G(Z) - \Gamma\big]q \qquad (3.27)$$

Here $d(z)$ is the group dispersion normalized by its average value.

Equation (3.27) may be reduced to a Hamiltonian structure by introducing a new amplitude $u = q/a$, where

$$\frac{da}{dZ} = \big[-\Gamma + G(Z)\big]a \qquad (3.28)$$

u then satisfies,

$$\frac{\partial u}{\partial Z} = i\frac{d(Z)}{2}\frac{\partial^2 u}{\partial T^2} + ia^2(Z)|u|^2 u \qquad (3.29)$$

Equation (3.29) is not integrable because of inhomogeneous coefficients $a(Z)$ and $d(Z)$.

The variational approach is one of the most powerful techniques capable of providing insight into the main characteristics of the DM soliton. The general idea underlying this approximation involves choosing an initial ansatz for the shape of solution sought, but leaves in the ansatz a set of free parameters, which may evolve with Z. An important feature of DM solitons is that their profile in general are not given by a *sech*, as it happens for the NLSE fundamental soliton, but varies from *sech* to a Gaussian function, and even more flat distribution. For the case of a generic structural function $Q(T)$, we may write

$$u(Z,T) = B(Z)Q(T)\exp\left\{i\varphi(Z) + i\frac{I(Z)}{T_p(Z)}T^2\right\} \quad (3.30)$$

where $B(Z)$ is the amplitude, $T_p(Z)$ is the width, $\phi(Z)$ is the phase, and $I(Z)$ is the normalized chirp parameter. Using the variational approach, we obtain the following evolution equations for $T_p(Z)$ and $I(Z)$:

$$\frac{dT_p}{dZ} = 2d(Z)I(Z) \quad (3.31)$$

$$\frac{dI}{dZ} = \frac{d(Z)K_1}{2T_p^3} - \frac{a^2(Z)K_2}{T_p^2} \quad (3.32)$$

The constants K_1 and K_2 are related to the structural function $Q(T)$. To describe periodic breathing pulses, solutions of these equations must satisfy two conditions of periodicity: $T(1) = T(0)$ and $I(1) = I(0)$. Some manipulations of Eqs. (3.31) and (3.32) give the following condition for the existence of the periodic solution:

$$\left\langle d(Z)/T_p^2 \right\rangle > 0 \quad (3.33)$$

It can be verified that the pulse width achieves a minimum value in the middle of each fiber segment, where the chirp becomes zero. However, the pulse width in the middle of the normal dispersion fiber is higher than in the middle of the anomalous dispersion fiber, which makes $\left\langle d(Z)/T_p^2 \right\rangle > 0$ even if the average dispersion is normal. In contrast, the linear stationary pulse is possible only if the average dispersion is zero This allows DM solitons to have much larger tolerance in the fiber dispersion.

Since a DM soliton can exist even when the average dispersion is zero, it is free from the Gordon–Haus jitter. Furthermore, the amplitude of the DM soliton can be made sufficiently large, by the so-called soliton energy enhancement factor, with a proper design of the dispersion map. That is, a DMS can be free from both amplitude and timing jitter and can be very robust in periodically amplified all-optical transmission systems.

The DM soliton is quite different than a standard fundamental soliton, since its shape, width, and peak power are not constant. However, since these parameters repeat from period to period at any location within the map, DM solitons can be used in optical communications. Actually, the remarkable characteristics of such solitons make them the preferred option for use in high-capacity transmission systems.

3.7 SOLITON-EFFECT COMPRESSION

Early work on optical pulse compression did not make use of any nonlinear optical effects. Only during the 1980s, after having understood the evolution of optical pulses in silica fibers, the self-phase modulation (SPM) effect was used to achieve pulse compression. Using such an approach, an optical pulse of 50-fs width from a colliding pulse-mode-locked dye laser oscillating at 620 nm was compressed down to 6 fs [70].

In one pulse compression scheme, known as *soliton-effect compressor*, the fiber itself acts as a compressor without the need of an external device [71,72]. The input pulse propagates in the anomalous-GVD regime of the fiber and is compressed through an interplay between SPM and GVD. This compression mechanism is related to a fundamental property of the higher-order solitons. As seen in Section 3.2, these solitons follow a periodic evolution pattern such that they go through an initial narrowing phase at the beginning of each period. If the fiber length is suitably chosen, the input pulses can be compressed by a factor that depends on the soliton order. Obviously, this compression technique is applicable only in the case of optical pulses with wavelengths exceeding 1.3 μm when propagating in standard single mode fibers (SMFs).

The evolution of a soliton of order N inside an optical fiber is governed by the NLSE, given by Eq. (3.5). Even though higher-order solitons follow an exact periodic evolution only for integer values of N, Eq. (3.5) can be used to describe pulse evolution for arbitrary values of N.

In practice, soliton effect compression is carried out by initially amplifying optical pulses up to the power level required for the formation of higher-order solitons. The peak optical power of the initial pulse required for the formation of the Nth-order soliton is given by Eq. (3.12).

These Nth-order solitons are then passed through the appropriate length of optical fiber to achieve a highly compressed pulse. The optimum fiber length, z_{opt}, and the optimum pulse compression factor, F_{opt}, of a soliton-effect compressor can be estimated from the following empirical relations [72]:

$$z_{opt} \approx z_0 \left(\frac{0.32}{N} + \frac{1.1}{N^2} \right) \tag{3.34}$$

$$F_{opt} \approx 4.1 N \tag{3.35}$$

where

$$z_0 = \frac{\pi}{2} L_D = \frac{\pi t_0^2}{2|\beta_2|} \tag{3.36}$$

is the soliton period in real units.

Compression factors as large as 500 have been attained using soliton-effect compressors [73]. However, the pulse quality is relatively poor, since the compressed pulse carries only a fraction of the input energy, while the remaining energy appears in the form of a broad pedestal. From a practical point of view, the pedestal is deleterious since it causes the compressed pulse to be unstable, making it unsuitable for some applications. Despite this, soliton-effect compressed pulses can still be useful because there are some techniques that can eliminate the pedestal. One of them consists in the use of a nonlinear optical loop mirror (NOLM) [74,75], which will be described in Chapter 7. Since the transmission of a NOLM is intensity dependent, the loop length can be chosen such that the low-intensity pedestal is reflected while the higher-intensity pulse peak is transmitted, resulting in a pedestal-free transmitted pulse.

One difficulty faced when using the soliton-effect compressor is that pulses with high peak power are required for the formation of high-order solitons in conventional fibers. The use of dispersion-shifted fibers (DSFs) with small values of β_2 at the operating wavelength can reduce the peak power required for soliton generation by an order of magnitude. However, because the soliton period z_0 is inversely proportional to $|\beta_2|$, longer lengths of DSFs will be required for solitons to achieve optimum compression. As a result, the total fiber loss experienced by those solitons will be larger, and the loss-induced pulse broadening will have a significant impact on the compressor global performance.

REFERENCES

1. A. Hasegawa and F. D. Tappert, *Appl. Phys. Lett.* **23**, 142 (1973).
2. L. F. Mollenauer, R. H. Stolen, and J. P. Gordon, *Phys. Rev. Lett.* **45**, 1095 (1980).
3. A. Hasegawa, *Opt. Lett.* **8**, 650 (1983).
4. J. P. Gordon and H. A. Haus, *Opt. Lett.* **11**, 665 (1986).
5. L. F. Mollenauer, M. J. Neubelt, M. Haner, E. Lichtman, S. G. Evangelides, and B. M. Nyman, *Electron. Lett.* **27**, 2055 (1991).
6. A. Mecozzi, J. D. Moores, H. A. Haus, and Y. Lai, *Opt. Lett.* **16**, 1841 (1991).
7. Y. Kodama and A. Hasegawa, *Opt. Lett.* **17**, 31 (1992).
8. L. F. Mollenauer, J. P. Gordon, and S. G. Evangelides, *Opt. Lett.* **17**, 1575 (1992).
9. L. F. Mollenauer, E. Lichtman, M. J. Neubelt, and G. T. Harvey, *Electron. Lett.* **29**, 910 (1993).
10. Y. Kodama and S. Wabnitz, *Opt. Lett.* **19**, 162 (1994).
11. M. Nakazawa, Y. Kamada, H. Kubota, and E. Suzuki, *Electron. Lett.* **27**, 1270 (1991).
12. Y. Kodama, M. Romagnoli, and S. Wabnitz, *Electron. Lett.* **28**, 1981 (1992).
13. M. Matsumoto, H. Ikeda, T. Uda, and A. Hasegawa, *J. Lightwave Technol.* **13**, 658 (1995).
14. M. F. Ferreira, M. V. Facão, and S. V. Latas, *Fiber Integrat. Opt.* **19**, 31 (2000).
15. N. Akhmediev and A. Ankiewicz, *Dissipative Solitons*; Springer, Berlin (2005).
16. A. A. Ankiewicz, N. N. Akhmediev, and N. Devine, *Optical Fiber Technol.* **13**, 91 (2007).
17. M. F. Ferreira and S. V. Latas, Dissipative Solitons in Optical Fiber Systems, in *Optical Fibers Research Advances*, F. Columbus (Ed.), Nova Science Publishers, Hauppauge, New York (2008).
18. N. J. Smith, F. M. Knox, N. J. Doran, K. J. Blow, and I. Bennion, *Electron. Lett.* **32**, 54 (1996).
19. W. Forysiak, F. M. Knox, and N. J. Doran, *Opt. Lett.* **19**, 174 (1994).
20. A. Hasegawa, S. Kumar, and Y. Kodama, *Opt. Lett.* **22**, 39 (1996).
21. M. Suzuki, I. Morita, N. Edagawa, S. Yamamoto, H. Toga, and S. Akiba, *Electron. Lett.* **31**, 2027 (1995).
22. A. Hasegawa (Ed.), *New Trends in Optical Soliton Transmission Systems*; Kluwer, Dordrecht, the Netherlands (1998).

23. C. S. Gardner, J. M. Greene, M. D. Kruskal, and R. M. Miura, *Phys. Rev. Lett.* **19**, 1095 (1967).
24. M. J. Ablowitz and P. A. Clarkson, *Solitons, Nonlinear Evolution Equations, and Inverse Scattering*; Cambridge University Press, New York (1991).
25. V. E. Zakharov and A. Shabat, *Sov. Phys. JETP* **34**, 62 (1972).
26. M. F. Ferreira, *Nonlinear Effects in Optical Fibers*; John Wiley & Sons, Hoboken, NJ (2011).
27. K. J. Blow and N. J. Doran, *Opt. Commun.* **52**, 367 (1985).
28. M. F. Ferreira, Optical Amplifiers, in *Encyclopedia of Optical Engineering*, R. Driggers (Ed.), Marcel Dekker, New York (2004).
29. M. F. Ferreira, *Opt. Commun.* **107**, 365 (1994).
30. J. P. Gordon, *Opt. Lett.* **11**, 662 (1986).
31. F. M. Mitschke and L. F. Mollenauer, *Opt. Lett.* **11**, 659 (1986).
32. M. V. Facão and M. F. Ferreira, *J.Nonlinear Math. Phys.* **8**, 112 (2001).
33. K. Smith and L. F. Mollenauer, *Opt. Lett.* **14**, 1284 (1989).
34. A. Hasegawa and Y. Kodama, *Solitons in Optical Communications*; Oxford University Press, Oxford, UK (1995).
35. M. F. Ferreira and S. V. Latas, *Optical Eng.* **41**, 1696 (2002).
36. N. N. Akhmediev, A. Ankiewicz and J. M. Soto-Crespo, *J. Opt. Soc. Am. B* **15**, 515 (1998).
37. N. N. Akhmediev, V. V. Afanasjev, and J. M. Soto-Crespo, *Phys. Rev. E* **53**, 1190 (1996).
38. J. M. Soto-Crespo, N. N. Akhmediev, and V. V. Afanasjev, *J. Opt. Soc. Am. B* **13**, 1439 (1996).
39. M. F. Ferreira, M. V. Facão, and S. V. Latas, *Photon. Optoelectron.* **5**, 147 (1999).
40. N. N. Akhmediev, A Rodrigues, and G. Townes, *Opt. Commun.* **187**, 419 (2001).
41. V. V. Afanasjev and N. N. Akhmediev, *Phys. Rev. E* **53**, 6471 (1996).
42. M. F. Ferreira and S. V. Latas, Bound States of Plain and Composite Pulses in Optical Transmission Lines and Fiber Lasers, in *Applications of Photonic Technology*, R. Lessard, G. Lampropoulos, and G. Schinn (Eds.), **4833**, p. 845, SPIE Proc., Bellingham (2002).
43. C. Normand and Y. Pomeau, *Rev. Mod. Phys.* **49**, 581 (1977).
44. P. Kolodner, *Phys. Rev. A* **44**, 6466 (1991).
45. R. Graham, *Fluctuations, Instabilities and Phase Transitions*, Springer, Berlin (1975).
46. N. Akhmediev and A. Ankiewicz, Dissipative Solitons in the Complex Ginzburg-Landau and Swift-Hohenberg Equations, in *Dissipative Solitons*, N. Akhmediev, A. Ankiewicz (Eds.), Springer, Heidelberg (2005).
47. N. Akhmediev, J. M. Soto-Crespo, M. Grapinet, Ph. Grelu, *Opt. Fiber Technol.* **11**, 209 (2005).
48. J. N. Kutz, Mode-Locking of Fiber Lasers via Nonlinear Mode-Coupling, in *Dissipative Solitons*, N. Akhmediev, A. Ankiewicz (Eds.), Springer, Heidelberg (2005).
49. U. Peschel, D. Michaelis, Z. Bakonyi, G. Onishchukov, and F. Lederer, Dynamics of Dissipative Temporal Solitons, in *Dissipative Solitons*, N. Akhmediev, A. Ankiewicz (Eds.), Springer, Heidelberg (2005).
50. N. Akhmediev and A. Ankiewicz, *Solitons, Nonlinear Pulses and Beams*, Chapman & Hall, London, UK (1997).
51. J. Soto-Crespo, N. Akhmediev, and A. Ankiewicz, *Phys. Rev. Lett.* **85**, 2937 (2000).
52. N. Akhmediev, J. Soto-Crespo, and G. Town, *Phys. Rev. E* **63**, 056602 (2001).
53. N. Akhmediev and A. Ankiewicz, *Dissipative Solitons*: "*From Optics to Biology and Medicine*," Lecture notes in Physics, Springer, Berlin (2008).
54. P. Grelu and N. Akhmediev, *Nat. Photon.* **6**, 84 (2012).
55. S. C. Latas, M. F. Ferreira, and A. S. Rodrigues, *Opt. Fiber Technol.* **11**, 292 (2005).

56. N. Akhmediev, J.-M. Soto-Crespo, and P. Grelu, *Phys. Lett. A* **372**, 3124 (2008).
57. W. Chang, A. Ankiewicz, J.-M. Soto-Crespo, and N. N. Akhmediev, *Phys. Rev. A* **78**, 023830 (2008).
58. W. Chang, J. Soto-Crespo, A. Ankiewicz, and N. Akhmediev, *Phys. Rev. A* **79**, 033840 (2009).
59. P. Grelu, W. Chang, A. Ankiewicz, J. Soto-Crespo, and N. Akhmediev, *J. Opt. Soc. Am. B* **27**, 2336 (2010).
60. A. Komarov, F. Amrani, A. Dmitriev, K. Komarov, and F. Sanchez, *Phys. Rev. A* **87**, 023838 (2013).
61. W. Du, H. Li, J. Li, Z. Wang, P. Wang, Z. Zhang, and Y. Liu, *Opt. Express* **27**, 8059 (2019).
62. S. C. Latas, M. F. S. Ferreira, and M. Facão, *J. Opt. Soc. Am. B* **35**, 1033 (2017).
63. S. C. Latas, M. F. S. Ferreira, *J. Opt. Soc. Am. B* **36**, 3016 (2019).
64. J. M. Soto-Crespo, N. N. Akhmediev, V. V. Afanasjev, and S. Wabnitz, *Phys. Rev. E* **55**, 4783 (1997).
65. V. V. Afanasjev, *Opt. Lett.* **20**, 704 (1995).
66. S. Cundiff, J. Soto-Crespo, and N. Akhmediev, *Phys. Rev. Lett.* **88**, 073903 (2002).
67. S. C. Latas and M. F. Ferreira, *Opt. Lett.* **35**, 1771 (2010).
68. S. C. Latas, M. V. Facão, and M. F. Ferreira, *Appl. Phys. B* **104**, 131 (2011).
69. S. C. Latas, M. F. Ferreira, and M. V. Facão, *Appl. Phys. B* **116**, 279 (2013).
70. R. L. Fork, C. H. Brito Cruz, P. C. Becker, and C. V. Shank, *Opt. Lett.* **12**, 483, (1987).
71. L. F. Mollenauer, R. H. Stolen, J. P. Gordon, W. J. Tomlinson, *Opt. Lett.* **8**, 289 (1983).
72. E. M. Dianov, Z. S. Nikonova, A. M. Prokhorov, and V. N. Serkin, *Sov. Tech. Phys. Lett.* **12**, 311 (1986).
73. A. S. Gouveia-Neto, A. S. L. Gomes, and J. R. Taylor, *J. Mod. Opt.* **35**, 7 (1988).
74. K. R. Tamura and M. Nakazawa, *IEEE Photon. Technol. Lett.* **11**, 230 (1999).
75. M. D. Pelusi, Y. Matsui, and A. Suzuki, *IEEE J. Quantum Electron.* **35**, 867 (1999).

4 Highly Nonlinear Fibers

4.1 INTRODUCTION

Conventional glass fibers for optical communications are made of fused silica and show an attenuation as low as 0.148 dB/km [1] with a broad low-loss optical window, which covers about 60 THz, ranging from 1260 to 1675 μm [2]. As seen in Chapter 2, a number of third-order nonlinear processes can occur in optical fibers [3]; these can grow to appreciable magnitudes over the long lengths available in fibers, even though the nonlinear index of the silica glass is very small ($n_2 = 2.7 \times 10^{-20}$ m^2/W) [4].

The nonlinear effects generated in the fibers severely affect the performance of optical communications systems, since they impose limits on the launched power of the signals, channel bit rate, channel spacing, transmission bandwidth, and hence the entire information capacity of such systems [5–7]. However, the same nonlinear effects have been positively utilized in various types of fiber lasers, such as Raman fiber lasers [8], Brillouin fiber lasers [9], parametric oscillators [10], and soliton-SRS wavelength tunable lasers [11]. In addition, due to femtosecond-order response time [12] and broad operation bandwidth [3], the fiber nonlinearities offer a variety of novel possibilities for optical amplification and all-optical processing, which are very promising for near-future high-capacity networks. With relatively low input powers, the combination of several nonlinear effects can generate an ultrabroad supercontinuum (SC), which finds applications in various areas of spectroscopy, optical coherence tomography, frequency metrology, optical communication sources, etc. [13].

In standard single-mode silica fibers, the nonlinear parameter has a typical value of only $\gamma \approx 1.3$ W^{-1}/km [14]. Such a value is too small for some applications, requiring highly efficient nonlinear processes, namely, for optical signal processing. As a consequence, a fiber length of several kilometers would generally be necessary in such applications. To shorten the length of interaction, highly nonlinear fibers (HNLFs) made with standard silica core and cladding compositions but with a smaller effective mode area, and hence a larger nonlinear coefficient ($\gamma \approx 10-20$ W^{-1}/km) [15] have been widely used. About 1 km of this type of fiber is usually used for the demonstration of different nonlinear effects.

The fiber nonlinearity can be further enhanced by appropriately tailoring its structure. Different types of silica-based microstructured optical fibers (MOFs) have been designed to address this purpose [16]. Fibers with a small core dimension and a cladding with a large air-fill fraction allow for extremely tight mode confinement, that is, small effective mode area, and hence, a higher value of γ. Using this approach, it has been possible to fabricate pure-silica MOFs with a nonlinear coefficient $\gamma \approx 70$ W^{-1}/km.

Significantly higher values of γ can be achieved by combining tight mode confinement with the use of glasses with a greater intrinsic material nonlinearity coefficient than that of silica. Examples of suitable glasses that have been used include bismuth oxide ($\gamma \approx 1100$ W^{-1}/km) [17], lead-silicate ($\gamma \approx 1860$ W^{-1}/km) [18], and chalcogenide ($\gamma \approx 2450$ W^{-1}/km) [19] glasses. In using such fibers, the required fiber length for several nonlinear processing applications was impressively reduced to the order of centimeters [3].

4.2 HIGHLY NONLINEAR SILICA FIBERS

The fiber nonlinearity is commonly characterized by the nonlinear parameter γ, which is given by [3]

$$\gamma = \frac{2\pi}{\lambda} \frac{n_2}{A_{\mathit{eff}}} \tag{4.1}$$

where λ is the light wavelength, n_2 is the nonlinear-index coefficient of the fiber core, and A_{eff} is the effective mode area. Equation (4.1) shows that for a fixed wavelength, since n_2 is determined by the material from which the fiber is made, the most practical way of increasing the nonlinear parameter γ is to reduce the effective mode area A_{eff}. However, the nonlinear parameter γ can also be enhanced using the dopant dependence of the nonlinear refractive index n_2. It has been shown that n_2 increases linearly with the relative index difference Δ_d for a GeO$_2$-doped bulk glass [20]. The relative index difference is defined as $\Delta_d = (n_d^2 - n_0^2)/2n_d^2$, where n_0 and n_d denote the refractive indexes of pure silica and doped glass, respectively. On the other hand, n_2 decreases linearly with Δ_d for a F-doped bulk glass.

In order to enhance the nonlinear parameter γ, the best option will be provided by a fiber heavily doped with GeO$_2$ in the core and having a large refractive index difference between the core and the cladding, because such fibers exhibit a large n_2 as well as a small effective area A_{eff} [21]. In typical HNLFs, the relative refractive index difference of the core to the outer cladding is around 3%, while the core diameter is around 4 μm. A W-cladding profile with a fluorine-doped depressed cladding having a refractive index lower than that of the outer cladding allows the single-mode operation in the communication bands as well as flexibility in designing of chromatic dispersion [22]. Using such approaches, new kinds of HNLFs with specific dispersive properties, namely dispersion-shifted fibers, dispersion-compensating fibers, dispersion-decreasing fibers, and dispersion-flattened fibers have been developed [23,24]. The values of the nonlinear parameter γ for most of these fibers are in the range 10–20 W^{-1}/km [24]. Increasing γ much above these values is not possible in these types of fibers, since the optical mode confinement is lost when the core diameter is further reduced.

4.3 TAPERED FIBERS

The fiber normalized frequency V is given by [3]

$$V = k_0 a n_1 \sqrt{2\Delta} \tag{4.2}$$

where k_0 is the vacuum wavenumber, a is the core radius, n_1 is the core refractive index, and

$$\Delta = (n_1 - n_c)/n_1 \tag{4.3}$$

is the relative index difference between the core and the cladding of the fiber, n_c being the cladding refractive index. A single-mode fiber is achieved when $V < 2.405$. If Δ is kept constant and the core radius a is reduced, the normalized frequency V decreases and the mode confinement is lost. Equation (4.2) shows that in order to maintain a given value of V when decreasing a, Δ must increase such that $a^2 \Delta$ remains constant.

The maximum value of Δ is achieved when the cladding material is replaced by air ($n_c \approx 1$). In such case, the mode remains confined to the core even if the core diameter is close to 1 µm. However, such confinement is reduced for smaller values of the core diameter.

Narrow-core fibers with air cladding have been produced by tapering standard optical fibers, with an original cladding diameter of 125 µm [25–27]. Tapering a fiber involves heating and stretching it to form a narrow waist connected to untapered fiber by taper transitions. If the transitions are gradual, light propagating along the fiber suffers very little loss.

The nonlinear parameter γ is significantly enhanced when the core diameter is reduced to values of the order of 2 µm. Considering the Gaussian approximation for the mode profile, and using $\lambda = 1$ µm for the wavelength of light transmitted through the fiber, $n_1 = 1.45$ for the silica core and $n_c = 1$ for the air cladding, it was found that a maximum value $\gamma \approx 370$ W^{-1}/km occurs for $V \approx 1.85$ [28].

Figure 4.1 shows the mode spot size as a function of the fiber diameter, considering different fiber materials: silica (SiO_2), a lead silicate glass ($SiO_2 - PbO$), and a bismuth silicate glass ($SiO_2 - Bi_{12}O_{18}$). We observe the existence of a minimum beam waist for a given value of the fiber diameter, which depends on the fiber material. Two regions are identified in Figure 4.1. Region I corresponds to the large evanescent-field region, which may be useful for sensing applications. Region II corresponds to the high-confinement region, which is more appropriate for nonlinear applications.

In recent years, there has been a great interest in tapered fibers with a core diameter below 1 µm [29–31]. Numerical results, obtained without using the Gaussian approximation, have shown that the mean-field diameter of the optical mode of such a nanoscale fiber with silica core and air cladding achieves a minimum value when the core diameter is 0.74λ [30]. The nonlinear parameter γ attains then a maximum value, which scales with λ^{-3}. Using $n_2 = 2.6 \times 10^{-20}$ m^2/W, a value $\gamma = 662$ W^{-1}/km is obtained at $\lambda = 0.8$ µm.

The dispersive properties of a tapered fiber are very sensitive to the core size and can be adjusted by changing the core diameter [30,32]. The zero-dispersion wavelength (ZDW) can be shifted to the visible range and, in some cases, a second ZDW appears at longer wavelengths, defining a wavelength window between the two ZDWs, which shows anomalous group velocity dispersion (GVD). This is illustrated in Figure 4.2, which shows the calculated dispersion spectra of taper waists for

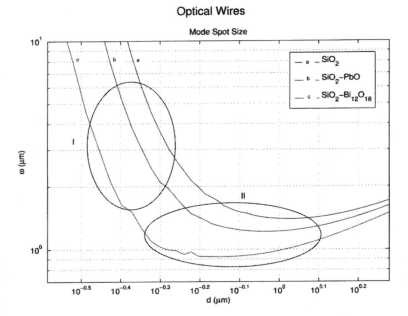

FIGURE 4.1 Variation of the spot size of a tapered fiber with the fiber diameter.

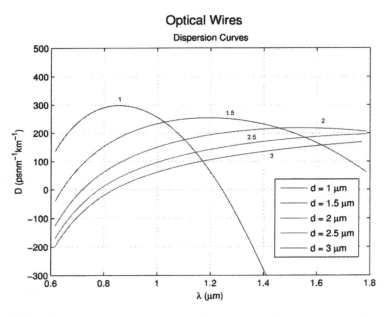

FIGURE 4.2 Dispersion properties of tapered fiber surround by air, for several values of the diameter.

several values of the waist diameter [32]. We observe that as the diameter of a taper waist decreases, the peak in dispersion shifts to shorter wavelengths and a second ZDW appears at longer wavelengths.

4.4 MICROSTRUCTURED FIBERS

MOFs represent a new class of optical fibers that are characterized by the fact that the silica cladding presents an array of embedded air holes. They are also referred as photonic crystal fibers because they were first realized in 1996 in the form of a photonic crystal cladding with a periodic array of air holes [33].

Microstructured optical fibers can be divided in two main types. One class of fibers, first proposed in 1999 [34], has a central region containing air. Such fibers are usually called hollow-core MOFs, and the light propagating in them is confined to the core by the photonic bandgap effect. The nonlinear effects are strongly reduced, and the dispersion becomes negligible in this kind of MOF. A nonlinear coefficient $\gamma = 0.023$ W^{-1}/km was reported for a hollow-core fiber [35]. However, the nonlinear effects in this type of MOFs can be greatly enhanced if air is replaced with a suitable gas or liquid [36].

The second type of microstructured fibers has a solid core, in which the light is guided mainly due to total internal reflection because the core has a higher refractive index than the cladding. In such fibers, the periodic nature of air holes in the cladding is not important as long as they provide an effective reduction of its refractive index below that of the silica core [37]. This helps to concentrate the mode field in a very small area, which is particularly the case in MOFs with small-scale cladding features and large air-filling fractions.

Figure 4.3 illustrates a solid-core MOF with an hexagonal pattern of holes in the cladding. This kind of fiber is characterized by two main structural parameters: the hair-hole diameter d and the hole-to-hole spacing Λ.

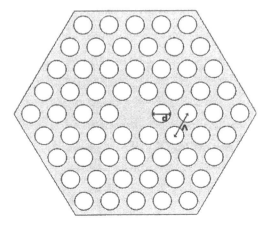

FIGURE 4.3 Schematic of a solid-core microstructured optical fiber with a hexagonal pattern of holes.

The number of guided modes in a conventional fiber is determined by the normalized frequency V given by Eq. (4.2), which depends on the wavelength. In conventional fibers, the single-mode condition $V < 2.405$ is only satisfied if the wavelength is in the infrared region. For lower values of the wavelength, the fiber becomes multimode. However, in a MOF, the normalized frequency is given by a slightly different expression from Eq. (4.2), with a replaced by Λ and n_c replaced by the effective cladding index of the microstructured cladding. Such effective refractive index depends significantly on the ratios d/Λ and λ/Λ [38] and asymptotically approaches the core index in the short wavelength limit. As a consequence, both the effective index difference and the numerical aperture go to zero in this limit.

Surprisingly, in a MOF it is verified that the behavior of the cladding-effective index cancels out the wavelength dependence of the normalized frequency V, such that it approaches a constant value as $\Lambda/\lambda \to \infty$. As a consequence, it was observed that a MOF shows a single mode behavior at all wavelengths if $d/\Lambda < 0.43$; such fiber is known as the *endlessly single-mode* (ESM) fiber [39,40]. Furthermore, because of the scale invariance of MOFs, the ESM behavior can be achieved for all core diameters.

The microstructured cladding offers greatly enhanced design flexibility and can manipulate the dispersion characteristics by controlling structural parameters, such as the hair-hole diameter d and the hole-to-hole spacing Λ [38]. In fact, the dispersive properties of MOFs are quite sensitive to these parameters and can be tailored by changing appropriately each of them. Figure 4.4 shows the dispersion characteristics of a microstructured fiber with a hexagonal pattern of holes spaced by $\Lambda = 2$ μm, considering a wavelength $\lambda = 1.55$ μm, for different hole-pitch ratios.

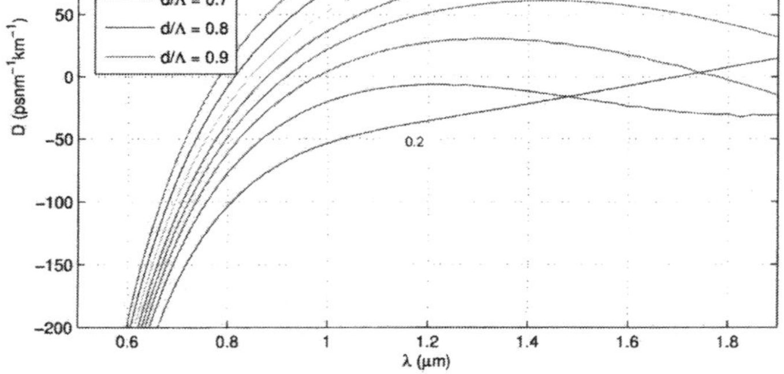

FIGURE 4.4 Dispersion properties of a microstructured fiber with a hexagonal pattern of holes spaced by $\Lambda = 2$ μm, considering a wavelength $\lambda = 1.55$ μm, for different hole-pitch ratios.

Highly Nonlinear Fibers

We observe that increasing the hole-pitch ratio, the ZDW can be shifted to the visible and near-infrared regions. In the case $d/\Lambda = 0.4$, two ZDWs are observed, defining a wavelength window with anomalous GVD. Both in this case and for $d/\Lambda = 0.3$, a nearly zero dispersion and flattened behavior are achieved. This is particularly important for telecom applications making use of the four-wave mixing effect.

If the core of a MOF is deliberately distorted so as to become twofold symmetric, extremely high values of birefringence can be achieved. For example, by introducing capillaries with different wall thicknesses above and below a solid glass core, values of birefringence some ten times larger than in conventional fibers can be obtained [41]. Moreover, experiments show that such birefringence is some 100 times less sensitive to temperature variations than in conventional fibers [42].

In a solid-core MOF, increasing the hole-pitch ratio reduces the effective mode area, due to the increased contrast between the refractive index of the core and the average refractive index of the cladding. Consequently, the nonlinear parameter is increased, as given by Eq. (4.1). Moreover, as the holes get larger, the core becomes more and more isolated, until it resembles an isolated strand of silica glass. The MOFs with larger cores exhibit semi-infinite anomalous dispersion above the ZDW. By decreasing the core size, the ZDW tends to be shifted to a shorter wavelength, leading to the anomalous dispersion at near infrared and visible wavelengths. When the core size is decreased further, a second ZDW arises in the longer wavelength side, such that the GVD is anomalous in the spectral window between the two ZDWs and normal outside it, as illustrated in Figure 4.4. Submicron-diameter MOF cores with low dispersion at 532 nm have been fabricated using a conventional tapering process [38].

In the case of microstructured fibers, and to take into account the different proportions of light in glass and air, the nonlinear coefficient γ must be redefined as follows [43]:

$$\gamma = k_0 \sum_i \frac{n_2^i}{A_{\textit{eff}}^i} = k_0 \frac{n_2^{\textit{eff}}}{A_{core}} \quad (4.4)$$

In Eq. (4.4), n_2^i is the nonlinear refractive index of material i (2.9×10^{-23} m^{2}W^{-1} for air, 2.6×10^{-20} m^{2}W^{-1} for silica); $A_{\textit{eff}}^i$ is the effective area for the light in material i; and $n_2^{\textit{eff}}$ is the effective nonlinear index for fiber, with a core area A_{core}. A nonlinear coefficient $\gamma = 240$ W^{-1}/km at 850 nm was measured for a solid-core MF with a core diameter of 1 μm [16], which must be compared with the highest values ~ 20 W^{-1}/km obtainable in conventional highly nonlinear silica fibers. Considering a peak power of $P_0 = 5$ kW, the nonlinear length for the same solid-core MOF will be $L_{NL} < 1$ mm. For dispersion values in the range $-400 < \beta_2 < 400$ ps^2/km and pulse durations $t_0 = 300$ fs, we have a dispersion length $L_D = t_0^2 / |\beta_2| > 0.2$ m. Considering also typical values of loss (1–100 dB/km), we conclude that both the effective fiber length $L_{\textit{eff}} = (1 - \exp(-\alpha L))/\alpha$ and the dispersion length L_D are much longer than the nonlinear length L_{NL}, which means that strong nonlinear effects will be readily observable in solid-core MOFs.

4.5 NON-SILICA FIBERS

The value of the nonlinear parameter γ can be enhanced if the optical fiber is made using glasses with higher nonlinearities than silica, such as lead silicate, tellurite, bismuth oxide, and chalcogenide glasses. Considering several optical glasses, it has been observed that the nonlinear-index coefficient n_2 increases with the linear refractive index, n [44,45]. Introducing heavy atoms or ions with a large ionic radius (i.e., using chalcogen elements S, Se, or Te to replace oxygen) act to increase the polarizability of the components in the glass matrix and also increase the nonlinear index n_2. Actually, the nonlinear index can be increased by a factor of almost 1000 compared with its value for silica fibers using some types of highly nonlinear glasses. The ZDW of a glass also shifts to longer wavelengths with increasing linear refractive index. Heavy metal oxide glasses (lead silicate, bismuth oxide, tellurite) have linear indices in the range 1.8–2.0, nonlinear indices ~10 times higher than silica, and material ZDWs of 2–3 μm. Chalcogenide glasses (GLS, As_2S_3) have linear indices of 2.2–2.4, nonlinear indices significantly higher than those of the oxide glasses, and material ZDWs larger than 4 μm.

Lead-silicate and tellurite glasses can offer values of n_2 which are about 10–20 times larger than that of silica. For example, a value $n_2 \approx 2.5 \times 10^{-19}$ m^2/W has been measured for tellurite glasses [46], while a value $n_2 = 4.1 \times 10^{-19}$ m^2/W at 1.06 μm was measured for Schott SF57 glass [18]. Both types of materials have been used for making microstructured fibers. The extrusion technique for fiber perform manufacture is particularly suitable when using compound glasses, since they exhibit low soft temperatures of ~500°C, as opposed to ~200°C for silica. Besides being simpler than the more common stacking technique developed for silica MOFs, extrusion allows the realization of cladding structures consisting of mostly air, as required to increase the fiber nonlinearity.

A nonlinear parameter $\gamma = 1860$ W^{-1}/km at 1.55 μm was measured in a 144-cm-long MOF made of lead-silicate glass (SF57) and presenting an effective mode area $A_{\mathit{eff}} = 1.1$ μm^2 [18]. Higher values of the nonlinear parameter are expected at shorter wavelengths due to a combination of the $1/\lambda$ dependence of γ and the fact that shorter wavelengths can be confined to the core down to smaller core dimensions.

Much attention has been paid in recent years to fibers based on bismuth oxide (Bi_2O_3) Among the different types of nonlinear fibers, Bi-NLFs have a relatively mature technology, and the fibers can be spliced to standard single-mode fibers. They also possess good chemical, mechanical, and thermal stability and have a low photosensitivity as compared to chalcogenide fibers. Depending on the concentration of bismuth in the fiber, the nonlinear refractive index n_2 can be varied from 30×10^{-20} m^2/W to 1.1×10^{-18} m^2/W, representing a 10–40 enhancement as compared with that of silica [17,47–49]. By 2008, a bismuth oxide fiber with a mode field diameter of 1.97 μm exhibited a nonlinear parameter $\gamma \approx 1100$ W^{-1}/km at 1550 nm. Only 35-cm length of such a highly nonlinear fiber was needed to demonstrate several applications on nonlinear processing of optical signals [17]. The large Brillouin gain offered by the Bi-NLF has found also important applications in slow light. A 2-m long Bi-NLF was successfully used to produce a fourfold reduction in the group velocity of 180-ns pulses using only 400 mW input power to provide the Brillouin gain [49].

A nonlinear parameter $\gamma = 2450$ W^{-1}/km was measured in a 2004 experiment for a 85-cm-long chalcogenide fiber with a core diameter of 7 μm [19]. This corresponds to a nonlinear refractive index $n_2 = 2.4 \times 10^{-17}$ m^2/W, which is larger by a factor of about 1000, relatively to the silica case. The same fiber also exhibited a Raman gain parameter, which was nearly 800 times larger than that of silica fibers [19]. This allows the use of either a smaller fiber length or a lower pump power to attain amplifier gain characteristics similar to those of silica-based devices. Due to their high nonlinearity, chalcogenide fibers have been considered as a good option to enhance several nonlinear effects.

In recent years, MOFs have been fabricated with various types of chalcogenide glasses for several applications. Employment of photonic crystal geometry in the cladding of As$_2$Se$_3$ fibers can give rise to a strong overlap between pump and signal, providing higher Raman gain efficiency than conventional fibers, as demonstrated for silica-based MOF structures [50,51]. It was shown that the Raman gain efficiency in As$_2$Se$_3$ MOFs can be improved by a factor of more than 4 compared with conventional As$_2$Se$_3$ fibers [52].

Compared with the bismuth fiber, a chalcogenide fiber has an intrinsically higher material nonlinearity (up to ×10 that of bismuth oxide). However, because the core area of available chalcogenide fibers is higher than that of bismuth fiber, the resulting nonlinear parameter γ is similar in both cases. It is expected that further progress in fabrication methods will lead to chalcogenide fibers with core areas comparable to the current bismuth fiber, which will provide nonlinear parameters $\gamma > 10,000$ W^{-1}/km [53].

Both the bismuth oxide fiber and the chalcogenide fiber present a relatively high normal dispersion at a 1550-nm communication window: typically −300 ps/nm/km in the first case and −500 ps/nm/km in the second one. Such large values can be useful in some applications but can cause limitations in other applications. In 2004, an air-cladding-type dispersion-shifted bismuth oxide MOFs has been successfully fabricated [54], which allows the control of dispersion characteristics.

4.6 SOLITON FISSION AND DISPERSIVE WAVES

As seen in Chapter 3, the propagation of femtosecond pulses in optical fibers can be perturbed by several higher-order dispersive and nonlinear effects, namely the third-order dispersion, self-steepening, and IRS. In the case of a higher-order soliton, these perturbations can determine the fission of the pulse into its constituent fundamental solitons, a phenomenon called *soliton fission*. This phenomenon was observed for the first time in a 1987 experiment, by observing the spectra at the output of a fiber with a 1 km-length of 830-fs input pulses with different values of the peak power [55].

In the presence of higher-order dispersion, an Nth-order soliton gives origin to N fundamental solitons whose widths and peak powers are given by [56]:

$$t_k = \frac{t_0}{2N+1-2k} \quad (4.5)$$

$$P_k = \frac{(2N+1-2k)^2}{N^2} P_0 \qquad (4.6)$$

where $k = 1$ to $\tilde{N}$, where $\tilde{N}$ is the integer closest to N when N is not an integer. Soliton fission occurs generally after a propagation distance $L_{fiss} \sim L_D/N$, where $L_D = t_0^2/|\beta_2|$ is the dispersion distance at which the injected higher-order soliton attains its maximum compression and larger bandwidth. The fission distance L_{fiss} is a particularly significant parameter in the context of higher-order soliton effect compression [57,58].

Besides higher-order dispersion, another main effect affecting the dynamics of a higher-order soliton fission is the intrapulse Raman scattering (IRS). As discussed in Chapter 3, this phenomenon leads to a continuous downshift of the soliton carrier frequency, an effect known as the *soliton self-frequency shift* (SSFS) [59]. Such effect was observed for the first time by Mitschke and Mollenauer in 1986 [60]. The origin of SSFS can be understood by noting that for ultrashort solitons, the pulse spectrum becomes so broad that the high-frequency components of the pulse can transfer energy through Raman amplification to the low-frequency components of the same pulse. Such an energy transfer appears as a red shift of the soliton spectrum, with shift increasing with distance.

The rate of frequency shift per propagation length is given by [3]

$$\frac{df}{dz} = -\frac{4t_R |\beta_2|}{15\pi t_0^4} = -\frac{4t_R (\gamma P_0)^2}{15\pi |\beta_2|} \qquad (4.7)$$

where $t_R \approx 5$ fs is the Raman parameter, and P_0 is the soliton peak power. Since the SSFS effect is proportional to $(\gamma P_0)^2$, it will be enhanced if short pulses with high peak power are propagated in highly nonlinear fibers.

Since its discovery in conventional single-mode fibers, the SSFS effect has been also observed in other types of fibers, including tapered fiber [61], solid-core MOFs [62], and air-core MOFs [63]. The SSFS effect can be significantly enhanced in some highly nonlinear fibers, where it has been used during the recent years for producing femtosecond pulses [64–67], as well as to realize several signal processing functions [68,69].

As a consequence of the SSFS induced by IRS, the solitons arising from the fission process separate from each other. Since the SSFS is the largest for the shortest soliton, its spectrum shifts the most toward the red side. In this process, new spectral peaks appear at $z = L_{fiss}$, corresponding to the so-called *nonsoliton radiation* (NSR) [70] or *Cherenkov radiation* [71], which is emitted by the solitons resulting from the fission process in the presence of higher-order dispersion. Such radiation was observed for the first time in a 2001 experiment using 110-fs pulses from a laser operating at 1556 nm, which were launched into a polarization-maintaining dispersion-shifted fiber with a GVD $\beta_2 = -0.1$ ps^2/km at the laser wavelength [72].

The NSR is emitted at a frequency such that its phase velocity matches that of the soliton. The two phases are equal when the following phase-matching condition is satisfied:

$$\beta(\omega) = \beta(\omega_s) + \beta_1(\omega - \omega_s) + \frac{1}{2}\gamma P_s \qquad (4.8)$$

Highly Nonlinear Fibers

In Eq. (4.8), P_s and ω_s are the peak power and the frequency, respectively, of the fundamental soliton formed after the fission process. Expanding $\beta(\omega)$ around ω_s, this phase-matching condition reduces to

$$\sum_{n\geq 2} \frac{\Omega^n}{n!} \beta_n(\omega_s) = \frac{\gamma P_s}{2} \tag{4.9}$$

where $\Omega = \omega - \omega_s$ is the frequency shift between the dispersive wave and the soliton, and β_n ($n = 2, 3$) are the dispersion parameters, calculated at the soliton central frequency ω_s.

Neglecting the fourth- and higher-order dispersion terms, the frequency shift between the frequency of dispersive waves and that of the soliton is given approximately by [71]:

$$\Omega_d \approx -\frac{3\beta_2}{\beta_3} + \frac{\gamma P_0 \beta_3}{3\beta_2^2} \tag{4.10}$$

For solitons propagating in the anomalous-GVD regime, we have $\beta_2 < 0$. In these circumstances, Eq. (4.13) shows that $\Omega > 0$ when $\beta_3 > 0$, in which case the NSR is emitted at wavelengths shorter than that of the soliton. On the contrary, if $\beta_3 < 0$, the NSR is emitted at a longer wavelength than that of the soliton, and it can result in the suppression of the SSFS effect [73,74].

4.7 FOUR-WAVE MIXING

As discussed in Chapter 2, new frequencies can be generated in conventional optical fibers by parametric four-wave mixing (FWM) when they are pumped close to the ZDW. In this case, the phase matching condition is sensitive to the exact shape of the dispersion curve. Thus, we must expect that the peculiar dispersive properties of some highly nonlinear fibers have a profound impact on the FWM process through the phase-matching condition.

The phase-matching condition for the FWM process requires that dispersive effects compensate nonlinear ones through the following equation:

$$\beta_2 \Omega_s^2 + \frac{\beta_4}{12} \Omega_s^4 + 2\gamma P_p = 0 \tag{4.11}$$

where β_2 and β_4 are, respectively, the second- and fourth-order dispersion terms at the pump frequency, Ω_s is the frequency shift between the signal or idler frequencies and the pump frequency, γ is the nonlinear parameter, and P_p is the pump peak power. The third and all-odd dispersion orders play no role in Eq. (4.11) since they cancel out from the degenerate phase-matching condition $2\beta(\omega_p) - \beta(+\Omega_s) - \beta(-\Omega_s) = 0$, where $\beta(\omega)$ is the exact wave vector.

The impact of the fourth-order dispersion term (β_4) on parametric processes was experimentally studied some years ago in standard telecommunications fibers [75] and in MOFs [76]. In the anomalous dispersion regime ($\beta_2 < 0$), the β_2 term dominates in Eq. (4.11), and the frequency shift is given by Eq. (2.47). However, modulation instability can also occur in the normal-GVD regime ($\beta_2 > 0$). In this case, the positive nonlinear phase mismatch $2\gamma P$ is compensated by the negative value of the linear phase mismatch due to β_4.

In general, the frequency shift between the pump and the FWM sidebands is much larger than the Raman gain band (of 13.2 THz) when the pump wavelength is below the ZDW. Therefore, no significant spectral overlap is possible between Raman and FWM gain bands in this situation. Such feature is particularly important to reduce the Raman-related noise and allows the use of the MF as a compact bright tunable single-mode source of entangled photon pairs, with wide applications in quantum communications [77–79].

FWM in highly nonlinear and microstructured optical fibers can be used to realize both fiber optic parametric amplifiers and optical parametric oscillators [80–84]. The physics of parametric amplification in HNLFs is similar to that of standard optical fibers. Differences arise, first, from the enhanced nonlinear parameter γ. Moreover, some MOFs can exhibit single transverse-mode propagation over a wide range of wavelengths, which leads to excellent spatial overlap between propagating modes at widely different wavelengths. Finally, the peculiar dispersive characteristics of HNLFs can enhance the phase-matching condition of the FWM process.

FWM in HNLFs can be used also to stabilize the output of multiwavelength erbium-doped fiber lasers, through the continuous annihilation and creation of photons at the wavelengths of interest [85]. A stable tunable dual-wavelength output over 21.4 nm was demonstrated using a 35-cm-long HNL bismuth-oxide fiber [86]. The number of oscillating wavelengths can be increased simply by applying a stronger FWM in the fiber. In such case, more wavelength components are involved in the energy exchange process to stabilize the laser output. Using the same 35-cm-long HNL bismuth-oxide fiber and a 30-dBm EDFA output power, up to four different wavelengths were produced in a 2008 experiment [17].

In addition to its use in generating continuous wave (CW) multiwavelength laser output, FWM in HNL bismuth-oxide fiber has also been used in several other applications, namely to produce wavelength and width-tunable optical pulses [87,88], frequency multiplication of a microwave photonic carrier [89], and wavelength conversion of 40 Gb/s polarization-multiplexed ASK-DPSK data signals [17].

REFERENCES

1. Y. Chigusa, Y. Yamamoto, T. Yokokawa, T. Sasaki, T. Taru. M. Hirano, M. Kakui, M. Onishi, and E. Sasaoka, *J. Lightwave Technol.* **23**, 3541 (2005).
2. Optical System Design and Engineering Consideration, ITU-T Recommendation Series G Supplement 39, February 2006.
3. M. F. Ferreira, *Nonlinear Effects in Optical Fibers*; John Wiley & Sons, Hoboken, NJ (2011).
4. N. G. Broderick, T. M. Monroe, P. J. Bennett, D. J. Richardson, *Opt. Lett.* **24**, 1395 (1999).
5. A. Chraplyvy, *J. Lightwave Technol.* **8**, 1548 (1999).
6. P. Mitra and J. Stark, *Nature* **411**, 1027 (2001).
7. M. Wu and W. Way, *J. Lightwave Technol.* **22**, 1483 (2004).

8. E. M. Dianov and A. M. Prokhorov, *IEEE J. Sel. Top. Quantum Electron.* **6**, 1022 (2000).
9. S. P. Smith, F. Zarinetchi, and S. Ezekiel, *Opt. Lett.* **16**, 393 (1991).
10. M. E. Marhic, K. K. Wong, L. G. Kazovsky, and T. E. Tsai, *Opt. Lett.* **27**, 1439 (2002).
11. N. Nishizawa and T. Goto, *J. Sel. Topics Quantum Electron.* **7**, 518 (2001).
12. E. M. Vogel, M. J. Weber, and D. M. Krol, *Phys. Chem. Glasses* **32**, 231 (1991).
13 J. Dudley and J. Taylor (Eds.), *Supercontinuum Generation in Optical Fibers*; Cambridge University Press, Cambridge, UK (2010).
14. T. Kato, Y. Suetsugu, M. Takagi, E. Sasaoka, and M. Nishimura, *Opt. Lett.* **20**, 988 (1995).
15. M. Hirano, T. Nakanishi, T. Okuno, and M. Onishi, *J. Sel. Topics Quantum Electron.* **15**, 103 (2009).
16. P. S. J. Russel, *J. Lightwave Technol.* **24**, 4729 (2006).
17. M. Fok and C. Shu, *IEEE J. Sel. Top. Quantum Electron.* **14**, 587 (2008).
18. J. Leong, P. Petropoulos, J. Price, H. Ebendorff-Heidepriem, S. Asimakis, R. Moore, K. Frampton, V. Finazzi, X. Feng, T. Monro, D. Richardson, *J. Lightwave Technol.* **24**, 183 (2006).
19. R. E. Slusher, G. Lenz, J. Hodelin, J. Sanghera, L. B. Shaw,. and I. D. Aggarwal, *J. Opt. Soc. Am. B* **21**, 1146 (2004).
20. K. Nakajima and M. Ohashi, *IEEE Photon. Technol. Lett.* **14**, 492 (2002).
21. T. Okuno, M. Onishi, T. Kashiwada, S. Ishikawa, and M. Nishimura, *IEEE J. Sel. Top. Quantum Electron.* **5**, 1385 (1999).
22. M. Takahashi, R. Sugizaki, J. Hiroishi, M. Tadakuma, Y. Taniguchi, and T. Yagi, *J. Lightwave Technol.* **23**, 3615 (2005).
23. K. Mori, H. Takara, and S. Kawanishi, *J. Opt. Soc. Am. B* **18**, 1780 (2001).
24. T. Okuno, M. Hirano, T. Kato, M. Shigemats, and M. Onishi, *Electron. Lett.* **39**, 972 (2003).
25. J. M. Harbold, F. O. Ilday, F. W. Wise, T. A. Birks, W. J. Wadsworth, and Z. Chen, *Opt. Lett.* **27**, 1558 (2000).
26. F. Lu and W. H. Knox, *Opt. Express* **12**, 347 (2004).
27. M. A. Foster and A. L. Gaeta, *Opt. Express* **12**, 3137 (2004).
28. A. Zheltitikov, *J. Opt. Soc. Am. B* **22**, 1100 (2005).
29. L. M. Tong, J. Y. Lou, and E. Mazur, *Opt. Express* **12**, 1025 (2004).
30. M. A. Foster, K. D. Moll, and A. L. Gaeta, *Opt. Express* **12**, 2880 (2004).
31. C. M. Cordeiro, W. J. Wadsworth, T. A. Birks, and P. St. J. Russel, *Opt. Lett.* **30**, 1980 (2005).
32. S. G. Leon-Saval, T. A. Birks, W. J. Wadsworth and P. StJ. Russel, *Opt. Express* **12**, 2864 (2004).
33. J. C. Knight, T. A. Birks, P. St. J. Russel, and D. M. Atkin, *Opt. Lett.* **21**, 1547 (1996).
34. R. F. Cregan, B. J. Mangan, J. C. Knight, *Science* **285**, 1537 (1999).
35. P. J. Roberts, F. Couny, H. Sabert, et al., *Opt. Express* **13**, 236 (2005).
36. P. Russell, *Science* **299**, 358 (2003).
37. B. J. Eggleton, C. Kerbage, P. S. Westbrook, R. S. Windeler, and A. Hale, *Opt. Express* **9**, 698 (2001).
38. K. Saito and M. Koshiba, *J. Lightwave Technol.* **23**, 3580 (2005).
39. T. Birks, J. Knight, and P. Russel, *Opt. Lett.* **22**, 961 (1997).
40. L. Dong, H. McKay, and L. Fu, *Opt. Lett.* **33**, 2440 (2008).
41. A. Ortigosa-Blanch, J. Knight, W. Wadsworth, J. Arriaga, B. Mangan, T. Birks, and P. Russel, *Opt. Lett.* **25**, 1325 (2000).
42. D. Kim and J. Kang, *Opt. Express* **12**, 4490 (2004).
43. J. Laegsgaard, N. Mortenson, J. Riishede, and A. Bjarklev, *J. Opt. Soc. Am. B* **20**, 2046 (2003).

44. X. Feng, A. K. Mairaj, D. W. Hewak, and T. M. Monro, *J. Lightwave Technol.* **23**, 2046 (2005).
45. J. H. V. Price, T. M. Monro, H. Ebendorff-Heidepriem, F. Poletti, P. Horak, V. Finazzi, J. Y. Y. Leong, P. Petropoulos, J. C. Flanagan, G. Brambilla, X. Feng, and D. J. Richardson, *IEEE J. Sel. Top. Quantum Electron.* **23**, 738 (2007).
46. J. S. Wang, E. M. Vogel, and E. Snitzer, *Opt. Mat.* **3**, 187 (1994).
47. J. H. Lee, K. Kikuchi, T. Nagashime, T. Hasegawa, S. Ohara, and N. Sugimoto, *Opt. Express* **13**, 3144 (2005).
48. J. H. Lee, T. Nagashima, T. Hasegawa, S. Ohara, N. Sugimoto, and K. Kikuchi, *J. Lightwave Technol.* **24**, 22 (2006).
49. C. Jauregui, H. Ono, P. Petropoulos, and D. J. Richardson, Proc. Opt. Fiber Commun. Conf. Mar. paper PDP2 (2006).
50. M. Fuochi, F. Poli, A. Cucinotta, and L. Vincetti, *J. Lightwave Technol.* **21**, 2247 (2003).
51. S. K. Varshney, T. Fujisawa, K. Saito, and M. Koshiba, *Opt. Express* **13**, 9516 (2005).
52. S. K. Varshney, K. Saito, K. Iizawa, Y. Tsuchida, M. Koshiba, and R. K. Sinha, *Opt. Lett.* **33**, 2431 (2008).
53. B. J. Eggleton, S. Radic, and D. J. Moss, Nonlinear Optics in Communications: From Crippling Impairments to Ultrafast Tools, in *Optical Fiber Telecommunications VA, Components and Subsystems*, I. Kaminoe, T. Li, and A. Willner (Eds.), Academic Press, San Diego, CA (2008).
54. H. Ebendorff-Heidepriem, P. Petropoulos, S. Asimadis, V. Finazzi, R. Moore, K. Frampton, F. Koizumi, D. Richardson, and T. Monro, *Opt. Express* **12**, 2082 (2004).
55. P. Beaud, W. Hodel, B. Zysset, and H. P. Weber, *IEEE J. Quantum Electron.* **23**, 1938 (1987).
56. Y. Kodama and A. Hasegawa, *IEEE J. Quantum Electron.* **23**, 510 (1987).
57. C. M. Chen and P. L. Kelley. *J. Opt. Soc. Am. B* **19**, 1961 (2002).
58. E. M. Dianov, Z. S. Nikonova, A. M. Prokhorov, and V. N. Serkin, *Pis'ma Zh. Tekh. Fiz.* **12**, 756 (1986) [Sov. Tech. Phys. Lett. **12**, 311–313 (1986)].
59. J. P. Gordon, *Opt. Lett.* **11**, 659 (1987).
60. F. Mitschke and L. Mollenauer, *Opt. Lett.* **11**, 659 (1986).
61. X. Liu, C. Xu, W. Knox, J. Chandalia, B. Eggleton, S. Kosinski, R. Windler, *Opt. Lett.* **26**, 358 (2001).
62. I. Cormack, D. Reid, W. Wadsworth, J. Knight, and P. Russel, *Electron. Lett.* **38**, 167 (2002).
63. D. Ouzounov, F. Ahmad, D. Muller, D. Venkataraman, M. Gallagher, M. Thomas, J. Silcox, K. Koch, and A. Gaeta, *Science* **301**, 1702 (2003).
64. N. Nishizawa, Y. Ito, T. Goto, *IEEE Photon. Technol. Lett.* **14**, 986 (2002).
65. A. Efimov, A. Taylor, F. Omenetto, and E. Vanin, *Opt. Lett.* **29**, 271 (2004).
66. K. Abedin and F. Kubota, *IEEE J. Sel. Top. Quantum Electron.* **10**, 1203 (2004).
67. J. Lee, J. Howe, C. Xu, X. and Liu, *J. Sel. Quantum Electron.* **14**, 713 (2008).
68. N. Nishizawa and T. Goto, *Opt. Express* **11**, 359 (2003).
69. S. Oda and A. Maruta, *Opt. Express* **14**, 7895 (2006).
70. A. Husakou and J. Herrmann, *Phys. Rev. Lett.* **87**, 203901 (2001).
71. N. Akhmediev and M. Karlsson, *Phys. Rev. A* **51**, 2602 (1995).
72. N. Nishizawa and T. Goto, *Opt. Express* **8**, 328 (2001).
73. D. V. Skryabin, F. Luan, J. C. Knight, and P. S. Russell, *Science* **301**, 1705 (2003).
74. F. Biancalana, D. V. Skryabin, and A. V. Yulin, *Phys. Rev. E* **70**, 016615 (2004).
75. S. Pitois and G. Millot, *Opt. Commun.* **226**, 415 (2003).
76. W. Wadsworth, N. Joly, P. Knight, T. Birks, F. Biancala, and P. Russel, *Opt. Express* **12**, 299 (2004).

77. J. G. Rarity, J. Fulconis, J. Duligall, W. J. Wadsworth, and P. St. J. Russell, *Opt. Express* **13**, 534 (2005).
78. J. Fan, A. Migdall and L. J. Wang, *Opt. Lett.* **30**, 3368 (2005).
79. J. Fulconis, O. Alibart, W. J. Wadsworth, P. St. J. Russell, and J. G. Rarity, *Opt. Express* **13**, 7572 (2005).
80. J. E. Sharping, *J. Lightwave Technol.* **26**, 2184 (2008).
81. J. E. Sharping, M. Fiorentino, P. Kumar, and R. S. Windeler, *Opt. Lett.* **27**, 1675 (2002).
82. Y. Deng, Q. Lin, F. Lu, G. Agrawal, and W. Knox, *Opt. Lett.* **30**, 1234 (2005).
83. J. E. Sharping, M. A. Foster, A. L. Gaeta, J. Lasri, O. Lyngnes, and K. Vogel, *Opt. Express* **15**, 1474 (2007).
84. J. E. Sharping, J. R. Sanborn, M. A. Foster, D. Broaddus, and A. L. Gaeta, *Opt. Express* **16**, 18050 (2008).
85. X. M. Liu, X. F. Yang, F. Y. Lu, J. H. Ng, X. Q. Zhou, and C. Lu, *Opt. Express* **13**, 142 (2005).
86. M. P. Fok and C. Shu, *Opt. Express* **15**, 5925 (2007).
87. C. Yu, L. S. Yan, T. Luo, Y. Wang, Z. Pan, and A. E. Wilner, *IEEE Photon. Technol. Lett.* **17**, 636 (2005).
88. M. P. Fok and C. Shu, *Proc. Fiber Commun. Conf., Mar.* paper OWI 34 (2006).
89. A. J. Seeds and K. J. Williams, *J. Lightwave Technol.* **24**, 4628 (2006).

5 Supercontinuum Generation

5.1 INTRODUCTION

One of the most impressive nonlinear phenomena that can be observed in highly nonlinear fibers is the supercontinuum generation (SCG). It corresponds to an extremely wide spectrum achieved by an optical pulse while propagating in a nonlinear medium, and results generally from the synergy between several fundamental nonlinear processes, such as self-phase modulation, cross-phase modulation, stimulated Raman scattering (SRS), and four-wave mixing (FWM) [1]. The spectral location and power of the pump, as well as the nonlinear and dispersive characteristics of the medium, determine the relative importance and the interaction between these nonlinear processes. The first observation of supercontinuum (SC) was realized in 1970 by Alfano and Shapiro in bulk borosilicate glass [2]. SC in fibers occurred for the first time in a 1976 experiment, when 10 ns pulses with more than 1-kW peak power were launched in a 20-m long fiber, producing a 180-nm-wide spectrum [3].

The SC generation can be broadly divided in two categories, depending on the duration of the pump pulses. The first category corresponds to long (picosecond, nanosecond, and continuous wave) pump pulses, whereas the second category is obtained with short (femtosecond) pump pulses. The dispersion regime (normal or anomalous) at the pump wavelength is also an important factor, at least in the initial propagation phase. Soliton-related effects play a significant role whenever light with sufficient power propagates in the anomalous dispersion regime [1].

In conventional dispersion-shifted fibers, the pulses are usually launched in the 1550 nm window, and the resulting SC spectrum extends at most from 1300 to 1700 nm. However, microstructure optical fibers (MOFs) can be designed with a zero-dispersion wavelength (ZDW) of anywhere from 1550 down to 565 nm [4]. Using such fibers, it becomes possible to generate a SC extending from the visible to the near infrared region. SCG has been achieved in different MOFs with pumping at 532 [5], 647 [6], 1064 [7], and 1550 nm [8].

A SC source can find applications in the area of biomedical optics, where it allows the improvement of longitudinal resolution in optical coherence tomography by more than an order of magnitude [9,10]; in optical frequency metrology [11,12]; in all kinds of spectroscopy, and as a multiwavelength source in the telecommunications area [13].

5.2 PUMPING WITH PICOSECOND PULSES

Pumping with long duration and high peak power pulses in the anomalous dispersion regime near the ZDW, the spontaneous modulation instability (MI) becomes the dominant process in the initial propagation phase. In the frequency domain, the

MI causes the spread of energy from the central wave frequency to sidebands with a frequency separation given by [14]:

$$\Omega_s = \sqrt{\frac{2\gamma P_p}{|\beta_2|}} \tag{5.1}$$

where β_2 is the group-velocity dispersion, γ is the nonlinear parameter, and P_p is the pump peak power. The MI period can be obtained from the above frequency separation as

$$T_s = \frac{2\pi}{\Omega_s} = \sqrt{\frac{2\pi^2 |\beta_2|}{\gamma P_p}} \tag{5.2}$$

The modulated picoseconds pulse can then break up into a train of solitons with durations approximately given by T_s. The subsequent evolution of these solitons then leads to additional spectral broadening and SC formation through a variety of mechanisms. If such solitons are sufficiently short, they are red-shifted due to the intrapulse Raman scattering effect [15,16]. As seen in Chapter 4, the rate of frequency shift is inversely proportional to the fourth power of the soliton width. If the solitons resulting from the initial pulse breakup are not sufficiently short, they can undergo soliton collisions, which determine the transfer of energy among them due to Raman scattering. In this process, some solitons achieve a higher peak power and then undergo a larger red-shift.

If the pump wavelength is near the ZDW, the red-shift of the solitons to longer wavelengths is accompanied by the generation of dispersive waves on the short-wavelength side of ZDW in fibers with positive dispersion slopes ($\beta_3 > 0$) [17]. Since the dispersive waves have a lower group velocity, they will lag behind the solitons. However, the group velocity of the red-shifting solitons decreases during the propagation, and eventually they meet with the dispersive waves. The temporal overlap between a soliton and a dispersive wave allows them to interact via the cross-phase modulation effect [14], which determines a blue-shift and a decrease in group velocity of the dispersive wave. In this way, as long as this cycle continues, the blue-shifted radiation cannot escape from the soliton, which is usually referred as a "trapping effect" [18,19]. This process is responsible for the maximum blue-shift of the SC, which is only limited by the ability of the solitons to red-shift. This red-shift can be limited by the existence of a second ZDW, which determines the spectral recoil of the soliton [20].

The solitonic effects are absent, at least in the initial propagation phase, if pumping is realized far into the normal dispersion regime. In this case, the spectral broadening occurs initially due to SRS. The contribution of the SRS process for the SCG is especially effective on the long-wavelength side of the spectrum, which tends to make the overall spectrum asymmetric.

The FWM becomes progressively more important if the pumping is closer to the ZDW, since the parametric gain is higher than the Raman gain [21]. Using pump pulses of subkilowatt peak power and a duration of some tens of picoseconds, a spatially single-mode SC more than 600 nm wide can be generated by the interplay of SRS and FWM [6,22].

The dispersive properties of the fiber become especially important in the case of the FWM process. If the required phase-matching conditions are satisfied, FWM generates sidebands on the short- and long-wavelength sides of the pulse spectrum. This process can produce a wideband SC even in the normal group velocity dispersion (GVD) region of an optical fiber. Actually, the phase-matching condition for the FWM process requires that dispersive effects compensate nonlinear ones through Eq. (4.11). This phase-matching condition can be satisfied even for a pump wavelength in the normal-GVD regime $(\beta_2 > 0)$ if the fourth-order dispersion coefficient (β_4) is negative. This can be easily realized in tapered and microstructured fibers [7,23].

In a 2008 experiment, a flat octave-spanning SC was generated from the visible to the near infrared by pumping a MOF at 1064.5 nm, corresponding to the normal dispersion regime, with 0.6-ns pulses with a 7 kHz repetition rate and ~60 mW average power [24]. Different fiber lengths were considered in this experiment. For a length $L = 1$ m, the signal and idler waves generated by degenerate FWM were clearly visible. They were located, respectively, at 810 and 1548 nm, which is in excellent agreement with Eq. (4.11). By increasing the fiber length, the spectral width of these two waves and that of the pump increased. The broadening of the pump spectrum was expected since the injected peak power is much higher than the Raman threshold. As a consequence, the phase-matching relation given by Eq. (4.11) could be satisfied between additional wavelengths located close to the pump and other wavelengths located around the signal and idler waves. For a fiber length of 30 m, the signal and idler waves started generating their own Raman cascade. For a fiber length of 100 m, the power of the signal wave became strong enough to generate up to five Raman Stokes orders located in the normal GVD region. In contrast, the Raman orders generated by the pump and idler waves were located in the anomalous regime and broadened rapidly, evolving into a continuum.

5.3 PUMPING WITH A CONTINUOUS WAVE

The general features observed in the case of long duration pulses can also be extended to the continuous wave (CW) pumping in the anomalous dispersion regime. In this case, the SCG also starts with the MI of the continuous pump wave in the anomalous dispersion region [25–28]. This instability evolves from background noise and is often stimulated by the intensity fluctuations, which are typical of a CW pump. The modulation's instability leads to the splitting of the CW into fundamental solitons.

In some circumstances, the fundamental solitons arising from MI can be short enough in order to enable the occurrence of intrapulse Raman scattering, leading to the soliton self-frequency shift [16,29]. As the MI evolves from noise, there will be some jitter on the period and hence also on the duration of the generated solitons. In this case, a range of soliton self-frequency shifts occurs, creating a smooth red-shifted continuum evolution.

Soliton collisions constitute also an important mechanism for the spectral expansion in CW pumped supercontinua [30,31]. In this case, the number of solitons can be particularly large and as red-shifted solitons travel more slowly than those at higher frequencies, they can eventually collide while propagating through a fiber. During such collisions, some energy will be transferred to the most red-shifted soliton, if it is within the spectral range of Raman scattering. The energy increase leads to a corresponding compression, which results in a further red-shift for that soliton, according with Eq. (4.7). The increased red-shift also slows down the soliton, leading it to pass by more solitons and hence experience more collision events. Soliton collisions can considerably increase the extent of a SC compared to Raman self-scattering alone.

The expansion of the SC to longer wavelengths can be limited by either the dispersion magnitude increasing or nonlinearity decreasing. Depending on the dispersion profile, the existence of a second ZDW can lead to the generation of dispersive waves and the spectral recoil of the solitons [20,32,33]. Concerning the extension to shorter wavelengths, it is determined by the excitation of phase-matched dispersive waves [17] and soliton-trapping mechanisms [18,19]. Experimentally, this phenomenon has been combined with dispersion-engineered MOFs to further extend SCs to the UV in nanosecond and picosecond pumping schemes [34]. The idea consists in modifying the dispersion curve the fiber in such a way that the group-velocity-matching conditions for trapped dispersive waves evolve continuously along propagation. This leads to the generation of new wavelengths as the ZDW decreases along propagation.

In a 2008 experiment, a SC ranging from 670 to 1350 nm with 9.55 W output power was achieved in the CW pumping regime, using a 200-m long ZDW decreasing MOF pumped by a 20 W ytterbium fiber laser at 1.06 μm [35]. The fiber was composed by a 100-m-long section with a constant dispersion followed by a 100-m-long section with linearly decreasing ZDW. To decrease the ZDW along propagation, the outer diameter of the second section fiber was linearly reduced from an initial value of 125 μm to a final value of 80 μm. The initial ZDW was located at 1053 nm, just below the pump wavelength and decreased to 950 nm in a quasi-linear way. The broadening of the output spectrum was shown to increase on both sides with increasing launch power. The spectrum was limited at long wavelengths by solitons red-shifted by the soliton self-frequency shift (SSFS) effect. On the other hand, the extension of the SC toward short wavelengths was mainly due to the trapping of dispersive waves by red-shifted solitons [19,36].

Compared with long-pulse-pumped systems, CW-pumped systems allow a simple experimental realization. Typically, the experimental arrangement consists simply of a CW fiber laser spliced directly to a suitable fiber. This brings advantages in terms of robustness and stability. An Er-doped fiber laser is generally used for pumping around 1400–1600 nm, whereas a Yb-doped fiber laser is commonly used around 1060 nm. Typically, a fiber-coupled power of 5–50 W is used for CW continuum generation. These relatively high values for the coupled power lead to very high spectral powers in the resulting continuum, typically greater than 10 mW/nm.

5.4 PUMPING WITH FEMTOSECOND PULSES

Pumping in the normal dispersion regime using femtosecond pulses, where soliton formation is not allowed, leads only to spectral broadening through self-phase modulation (SPM) [37,38]. As seen in Chapter 2, SPM-induced spectral broadening depends on the propagation distance, peak power, and shape of the input pulse. Its role becomes particularly significant at high input powers and for ultrashort pulses, so that the rate of power variation with time, dP/dt, is large. If the pump wavelength is located near the ZDW, the spectrum can extend into the anomalous dispersion regime. In such case, the soliton dynamics described above can also have a significant contribution to the spectral broadening [39].

Soliton dynamical effects play a prominent role when pumping in the anomalous dispersion regime using femtosecond pulses. If the power of the pump pulses is high enough, they can evolve as higher-order solitons. As seen in Chapter 3, these pulses begin experiencing a spectral broadening and temporal compression, which are typical of higher-order solitons [14]. However, because of perturbations such as higher-order dispersion or intrapulse Raman scattering, the dynamics of such pulses departs from the behavior expected of ideal high-order solitons, and the pulses break up. This process is known as soliton fission and determines the separation of each higher-order soliton pulse into N fundamental solitons whose widths and peak powers are given by Eqs. (4.5) and (4.6) [40].

The soliton fission process is affected if the input pulse wavelength is near a ZDW. As seen in Chapter 4, in such case the ejected solitons shed some of their energy to dispersive waves generated in the normal dispersion regime, the so-called nonsoliton radiation [17,41]. In spite of the fact that Raman solitons and dispersive waves have widely separated spectra, they can overlap in the time domain. In such case, they interact through cross-phase modulation (XPM), causing the formation of new spectral components. This results in the spectral broadening of the dispersive waves and merging of different peaks, helping in the formation of a broadband SC. This shows that the XPM effect plays indeed an important role during the SC formation in highly nonlinear fibers [42–45].

The group-velocity mismatch between the soliton and the dispersive wave can be relatively small if this wave propagates in the normal-dispersion regime. In such case, if they overlap in the time domain, the Raman soliton can trap the dispersive wave due to the XPM interaction [44,45]. The condition for this to occur can be obtained from the relation

$$\beta_1(\omega_d) \approx \beta_1(\omega_s) + \beta_2(\omega_s)\Omega_d + \frac{1}{2}\beta_3(\omega_s)\Omega_d^2 \qquad (5.3)$$

where $\Omega_d = \omega_d - \omega_s$ is the difference between the frequency of dispersive waves (ω_d) and that of the soliton (ω_s). Considering that $v_g = 1/\beta_1$, we have $\beta_1(\omega_d) \approx \beta_1(\omega_s)$ and the dispersive wave can be trapped by the soliton if

$$\Omega_d = -\frac{2\beta_2}{\beta_3} \qquad (5.4)$$

The FWM effect also plays an important role in the process of SC formation by femtosecond pulse pumping in the anomalous dispersion regime. Provided an appropriate phase-matching condition is satisfied, a Raman soliton can act as a pump and interact through FWM with a dispersive wave, giving origin to an idler wave, as described in Section 2.6 [38,46]. In one possible FWM process, corresponding to the conventional FWM, the phase-matching condition is given by

$$\beta_d(\omega_3) + \beta_d(\omega_4) = \beta_s(\omega_3) + \beta_s(\omega_4) \tag{5.5}$$

where $\beta_d(\omega)$ and $\beta_s(\omega)$ are the propagation constants of the dispersive wave and of the soliton, respectively. In this case, the idler frequency ω_4 lies far from the original dispersive wave. Considering a positive third-order dispersion, this FWM process contributes to the extension of the SC toward the infrared side of the input wavelength.

Another possibility corresponds to a Bragg-scattering-type FWM process, in which the phase-matching condition is given by

$$\beta_d(\omega_3) - \beta_d(\omega_4) = \beta_s(\omega_3) - \beta_s(\omega_4) \tag{5.6}$$

In this case, the idler frequency ω_4 lies close to the original dispersive wave. Considering a positive third-order dispersion, this FWM process plays an important role in extending the SC toward the blue side of the input wavelength.

Modeling the ultrashort pulse propagation and SCG in optical fibers can be realized considering a generalized nonlinear Schrödinger equation (NLSE) that includes higher-order dispersive and nonlinear effects [47–51]. Such equation can be written as:

$$\frac{\partial U}{\partial z} - i \sum_{k \geq 2} \frac{i^k \beta_k}{k!} \frac{\partial^k U}{\partial \tau^k} + \frac{\alpha(\omega)}{2} U = i\gamma \left(1 + \frac{i}{\omega_0} \frac{\partial}{\partial \tau}\right)$$
$$\left(U(z,\tau) \int_{-\infty}^{+\tau} R(t') |U(z,\tau-t')|^2 dt'\right) \tag{5.7}$$

where $U(z,t)$ is the electric field envelope, ω_0 is the center frequency, β_k are the dispersion coefficients at the center frequency, $\alpha(\omega)$ is the frequency-dependent fiber loss, and γ is the nonlinear parameter of the fiber.

The nonlinear response function $R(t)$ in Eq. (5.7) can be written as

$$R(t) = (1 - f_R)\delta(t) + f_R h(t), \tag{5.8}$$

where the δ-function represents the instantaneous electron response (responsible for the Kerr effect), $h(t)$ represents the delayed ionic response (responsible for the Raman scattering), and f_R is the fractional contribution of the delayed Raman response to the nonlinear polarization, in which a value $f_R = 0.18$ is often assumed [52]. It is common to approximate $h(t)$ in the form [53,54]:

$$h(t) = \frac{\tau_1^2 + \tau_2^2}{\tau_1 \tau_2^2} \exp(-t/\tau_2) \sin(t/\tau_1) \tag{5.9}$$

where $\tau_1 = 12.2$ fs and $\tau_2 = 32$ fs. More accurate forms of the response function $h(t)$ have also recently been investigated [55]. The Raman parameter, t_R, which characterizes the intrapulse Raman scattering effect is related with $h(t)$ in the form $t_R \equiv f_R \int_{-\infty}^{+\infty} th(t)dt$. Typically, it has a value $t_R \approx 5$ fs.

Supercontinuum Generation

Equation (5.7) can be used to describe the propagation of femtosecond pulses in optical fibers, in both the normal and anomalous dispersion regimes. When such pulses have enough power, their spectra undergo extreme broadening. In the anomalous dispersion regime, this process is mainly influenced by the phenomenon of soliton fission, which occurs whenever a higher-order soliton is affected by third- or higher-order dispersion. The soliton order is given by $N = \sqrt{L_D / L_{NL}}$, where $L_D = t_0^2 / |\beta_2|$ is the dispersion distance and $L_{NL} = 1/\gamma P_0$ is the nonlinear length, as defined in Chapter 2.

Figure 5.1 shows the temporal and spectral evolution of an optical pulse propagating along a silica microstructured optical fiber with a hexagonal pattern of holes of diameter $d = 1.4$ μm spaced by $\Lambda = 1.6$ μm. This result was obtained taking into account the full dispersion curve of such fiber, which presents a single ZDW at 735 nm. The pump is in the anomalous dispersion regime at 790 nm, where the nonlinear parameter is $\gamma = 117 \text{W}^{-1}\text{km}^{-1}$. An input pulse $U(0,\tau) = \sqrt{P_0}\text{sech}(\tau/t_0)$, is assumed, where $P_0 = 5$ kW, and the width $t_0 = 14.2$ fs, which corresponds to an intensity full width at half maximum (FWHM) of 25 fs. For the assumed pulse and fiber parameters, the soliton order is $N = 3.01$.

For the pulse and fiber parameters assumed, we have a fission distance $L_{fiss} \sim 0.51$ cm, which can be confirmed in Figure 5.1. A clear signature of soliton fission is the appearance of a new spectral peak in Figure 5.1a at $z = L_{fiss}$, which corresponds to the nonsoliton radiation.

As a consequence of the SSFS induced by intrapulse Raman scattering, the solitons arising from the fission process separate from each other. Since the SSFS is the largest for the shortest soliton, its spectrum shifts the most toward the red side

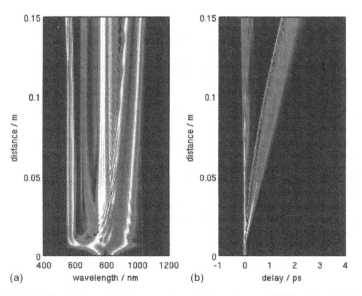

FIGURE 5.1 (a) Spectral and (b) temporal evolution along a MOF with $d = 1.4$ μm and $\Lambda = 1.6$ μm. The input pulse has a peak power of 5 kW and a FWHM of 25 fs.

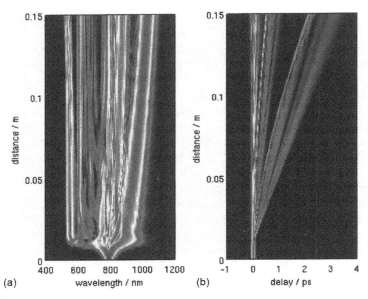

FIGURE 5.2 (a) Spectral and (b) temporal evolution of an optical pulse along a MOF with $d = 1.4$ μm and $\Lambda = 1.6$ μm. The input pulse has a peak power of 5 kW and a FWHM of 50 fs.

in Figure 5.1a. The change of the soliton's frequency determines a reduction in the soliton's speed because of dispersion. This deceleration appears as a bending of the soliton trajectory in the time domain, as observed in Figure 5.1b.

Figure 5.2 shows the temporal and spectral evolution of an optical pulse for the same conditions of Figure 5.1, except that the pulse width is now $T_0 = 28.4$ fs, which corresponds to an intensity FWHM of 50 fs. Comparing with Figure 5.1, we have now a wider and more uniform SC. Figure 5.3 shows the initial and final pulse profiles, both in frequency and time domains. Since the power of the first ejected soliton is greater than in the case of Figure 5.1, it experiences a larger SSFS, according with Eq. (4.7). At the same time, the corresponding nonsolitonic radiation is generated at a lower wavelength in the normal region.

In most SCG studies, silica has been used as the MOFs' background material. However, the material loss in silica increases drastically at wavelengths beyond ~2 μm, effectively preventing the spectral evolution of SCG into the mid-infrared (mid-IR) region [56].

Research into other highly nonlinear materials that can expose the importance of the mid-IR region has been increased in recent years. Price et al. [47] have theoretically generated a mid-IR SC from 2 to 5 μm using a bismuth-glass MOF. Domachuk et al. [57] have experimentally generated a mid-IR SC with a spectral range of 0.8–4.9 μm using a tellurite MOF with the same structure. Fluoride, sapphire, As_2Se_3, and $As_{40}Se_{60}$ chalcogenide MOFs have also been used to achieve mid-infrared SCG [58,59]. A simulated SC extending from 500 nm to 3900 nm, achieved after a propagation distance of only 0.3 mm, has been recently reported [60].

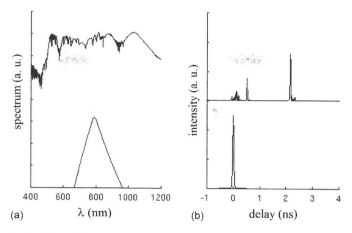

FIGURE 5.3 Initial and final pulse profiles in (a) frequency and (b) time domains for the case of Figure 5.2.

5.5 SUPERCONTINUUM COHERENCE

The coherence of the SC can be characterized by the degree of coherence associated with each spectral component, given by [48,61]

$$g_{12}(\omega,z) = \left| \frac{\left\langle \tilde{U}_1^*(\omega,z)\tilde{U}_2(\omega,z) \right\rangle}{\left[\left\langle \left|\tilde{U}_1(\omega,z)\right|^2 \right\rangle \left\langle \left|\tilde{U}_2(\omega,z)\right|^2 \right\rangle \right]^{1/2}} \right| \quad (5.10)$$

where $\tilde{U}_1$ and $\tilde{U}_2$ denote two outputs resulting from two independently generated inputs with random noise, and the angle brackets denote an average calculation on a set of pairs. It has been shown that $g_{12}(\omega)$ depends on the average value of the input pulse parameters, namely on the pulse width [48].

The SC spectra generated with subpicosecond pulses with normal dispersion regime pumping are expected to be always highly coherent, since MI does not occur in the normal GVD regime. However, the drawback is that the SC spectral width (at the same peak power) is comparatively much smaller due to the rapid initial temporal spreading of the pump pulses. Pumping in the anomalous dispersion regime, a higher coherence is expected for SC generated with shorter input pulses, where self-phase modulation plays a more significant role in spectral broadening and the effects of MI are less significant. This was confirmed in Refs. [48,62], which compared simulation results for 50, 100, and 150 fs input pulses, while maintaining the peak power and wavelength at 10 kW and 850 nm in the fiber anomalous dispersion regime. Although the spectral broadening in all cases was comparable, it was observed that the coherence properties improved significantly as the input pulse duration was decreased. Actually, it was verified that $g_{12}(\omega) \approx 1$ over more than an octave for the shortest input pulses of 50 fs. Moreover, the results of simulations for the case of 150-fs input pulses at an input wavelength of 740 nm in the normal dispersion regime, where MI is completely inhibited, have shown a negligible coherence degradation.

In the case of shorter pulses launched in the anomalous dispersion regime, they get temporally compressed so quickly that the soliton fission process dominates the initial dynamic. In this case, the spectral extent of the developing SC can overlap with the frequencies of maximum MI gain before significant amplification of the noise background occurred. However, in the case of longer pulses, MI-amplified noise becomes a dominant feature. Such amplified noise is random and can induce soliton fission even before other perturbations such as higher-order dispersive and nonlinear effects become significant. In this case, soliton fission occurs randomly, and fission products are completely incoherent. For typical experimental conditions, it has been found that input pulses where $N < 20$ possess high coherence, and input pulses where $N > 40$ possess low coherence [48,62].

As seen in Section 5.3, SCG does not involve the fission process when a CW pump is used in the anomalous dispersion region. In this case, SCG starts with the MI process, which leads to the splitting of the CW into a multitude of fundamental solitons. As the MI evolves from noise, there will be some jitter on the period and hence also on the duration of the generated solitons. Several studies have shown that, in this case, the incoherence of the SC results through a thermalization process leading to the formation of spectrally incoherent solitons [63–65]. Spectrally incoherent solitons were found numerically in 2008 [66] and observed experimentally for the first time in 2009 [67].

5.6 THE SUPERCONTINUUM AS A SOURCE FOR WDM SYSTEMS

Actual wavelength division multiplexing (WDM) transmission systems use a large number of distributed feedback (DFB) lasers as transmitters. Each DFB laser corresponds to a wavelength channel and is generally used together with a modulator and a wavelength-stabilizing system. This approach is costly and becomes impractical when the number of channels becomes large.

One approach to realize a practical WDM source uses the technique of spectrum slicing. In this approach, the light of a broadband source is sliced into several narrow bands centered at the wavelengths of the WDM channels. A multipeak optical filter can be used to realize the spectrum slicing. Since some of the components are shared among all the wavelengths, the number of active elements in such multiwavelength transmitters is greatly reduced.

The spectral width of femtosecond pulses emitted by a mode-locked laser is already sufficiently large to begin with and can be used for frequency slicing. In fact, according to the Fourier theorem for bandwidth-limited pulses, the temporal duration and the bandwidth are in inverse proportion to each other. For example, a laser producing a train of 50-fs transform-limited Gaussian pulses has an optical bandwidth of 70 nm at 1550 nm. The bandwidth of such pulses can be further broadened by chirping them using 10–15 km of a standard telecommunication fiber.

Spectrum slicing for WDM applications has been reported employing amplified spontaneous emission from light-emitting diodes [68,69], superluminescent diodes [70,71], or erbium-doped fiber amplifiers [72]. However, the transmission capacity of such incoherent spectrum-sliced systems is generally limited by the spontaneous emission beat noise. This limitation can be alleviated using coherent light sources, like mode-locked lasers [73,74] and SC generators [75–78]. The coherent nature of the broadband radiation originating from these sources is crucial to achieve high transmission capacities both in multiwavelength optical time division multiplexing and in dense wavelength division multiplexing (DWDM) systems.

Three characteristics of the SC are particularly important in the context of its application as a WDM system transmitter. First, its spectrum should ideally coincide with the transmission window. Second, it should be smooth enough, in order to provide similar powers to each channel, resulting in uniform impact of the nonlinear effects. Third, the amplitude noise of the filtered SC should be sufficiently reduced. This last characteristic becomes especially important for high-capacity WDM systems, since the intensity variations among pulses at a given wavelength has a negative impact on the performance of that channel.

Early in 1995, 3.5-ps pulses from a mode-locked fiber laser were broadened spectrally up to 200 nm through SCG by exploiting the nonlinear effects in a 3-km long optical fiber, resulting in a 200-channel WDM source [79]. Two years later, a SC source was used to demonstrate data transmission at a bit rate of 1.4 Tb/s using seven WDM channels, each operating at 200 Gb/s, with 600 GHz spacing [80]. By 2002, a 200-nm-wide SC was used to create a WDM source with 4200 channels with only 5 GHz spacing [81]. Such WDM source was latter used to create 50-GHz-spaced channels in a spectral range between 1425 to 1675 nm [82].

The structure of the cavity modes present in the original laser output is preserved when a normal dispersion fiber is used to generate the SC [83]. This enables the generation of an ultra-broadband frequency comb, in which the separation between peaks corresponds to the microwave mode-locking frequency of the source laser. Each peak can be considered then as a potential transmission channel. This property was used by Takara et al. [84] to generate more than 1000 optical frequency channels with a channel spacing of 12.5 GHz between 1500 and 1600 nm. Following the same principle, 124 nm seamless transmission of 3.13 Tb/s (10-channel DWDM × 313 Gb/s) over 160 km, with a channel spacing of 50 GHz, was reported [85]. Raman amplification in hybrid tellurite/silica fiber was used in this experiment to improve gain flatness. More recently, a field demonstration of 1046 channel ultra-DWDM transmission over 126 km was realized using a SC multicarrier source spanning 1.54–1.6 nm, which was mainly generated through SPM-induced spectral broadening. The channel spacing was 6.25 GHz and the signal data rate per channel 2.67 Gb/s [13]. The conservation of coherence properties was also successfully employed by Sotobayashi et al. [85] to create a 3.24 Tb/s (84-channel 40 Gbit/s) WDM source of carrier-suppressed return-to-zero pulses.

REFERENCES

1. J. Dudley and J. Taylor (Eds.), *Supercontinuum Generation in Optical Fibers*; Cambridge University Press, Cambridge, UK (2010).
2. R. R. Alfano and S. L. Shapiro, *Phys. Rev. Lett.* **24**, 584 (1970).

3. C. Lin and R. H. Stolen, *Appl. Phys. Lett.* **28**, 216 (1976).
4. W. H. Reeves, D. V. Skryabin, F. Biancalana, J. C. Knight, P. St. J. Russel, F. G. Omenetto, A. Efomov, and A. J. Taylor, *Nature* **424**, 511 (2003).
5. S. G. Leon-Saval, T. A. Birks, W. J. Wadsworth, P. S. J. Russel, and M. W. Mason, *Opt. Express* **12**, 2864 (2004).
6. S. Coen, A. H. L. Chau, R. Leonhardt, J. D. Harvey, J. C. Knight, W. J. Wadsworth, and P. S. J. Russell, *J. Opt. Soc. Am. B* **19**, 753 (2002).
7. W. J. Wadsworth, N. Joly, J. C. Knight, T. A. Birks, F. Biancalana, and P. J. L. Russell, *Opt. Express* **12**, 299 (2004).
8. V. V. Ravi, K. Kumar, A. K. George, W. H. Reeves, J. C. Knight, P. S. J. Russell, F. G. Omenetto, and A. J. Taylor, *Opt. Express* **10**, 1520 (2002).
9. A. F. Fercher, W. Drexler, C. K. Hitzenberger, and T. Lasser, *Rep. Progr. Phys.* **66**, 239 (2003).
10. G. Humbert, W. J. Wadsworth, S. G. Leon-Saval, J. C. Knight, T. A. Birks, P. S. J. Russell, M. J. Lederer, D. Kopf, K. Wiesauer, E. I. Breuer, and D. Stifter, *Opt. Express* **14**, 1596 (2006).
11. M. Bellini and T. W. Hänsch, *Opt. Lett.* **25**, 1049 (2000).
12. H. Hundertmark, D. Wandt, C. Fallnich, N. Haverkamp, and H. Telle, *Opt. Express* **12**, 770 (2004).
13. H. Takara, T. Ohara, T. Yamamoto, H. Masuda, M. Abe, H. Takahashi, and T. Morioka, *Electron. Lett.* **41**, 270 (2005).
14. M. F. Ferreira, *Nonlinear Effects in Optical Fibers*; John Wiley & Sons, Hoboken, NJ (2011).
15. F. M. Mitschke and L. F. Mollenauer, *Opt. Lett.* **11**, 659 (1986).
16. J. P. Gordon, *Opt. Lett.* **11**, 662 (1986).
17. N. Akhmediev and M. Karlsson, *Phys. Rev. A* **51**, 2602 (1995).
18. N. Nishizawa and T. Goto, *Opt. Express* **10**, 1151 (2002).
19. A. V. Gorbach and D. V. Skryabin, *Nat. Photonics* **11**, 653 (2007).
20. D. W. Skryabin., F. Luan., J. C. Knight, and P. S. J. Russell, *Science* **301**, 1705 (2003).
21. G. P. Agrawal, *Nonlinear Fiber Optics*, 4th ed., Academic Press, Burlington, MA, (2007).
22. J. M. Dudley, L. Provino, N. Grossard, H. Maillotte, R. S. Windeler, B. J. Eggleton, and S. Coen, *J. Opt. Soc. Am. B* **19**, 765 (2002).
23. G. K. Wong, A. Y. Chen, S. G. Murdoch, R. Leonhardt, J. D. Harvey, N. Y. Joly, J. C. Knight, W. J. Wadsworth, and P. St. J. Russel, *J. Opt. Soc. Am. B* **22**, 2505 (2005).
24. A. Kudlinski, V. Pureur, G. Bouwrnans, and A. Mussot, *Opt. Lett.* **33**, 2488 (2008).
25. A. Hasegawa and W. F. Brinkman, *IEEE J. Quantum Electron.* **16**, 694 (1980).
26. H. Itoh, G. M. Davis, and S. Sudo, *Opt. Lett.* **14**, 1368 (1989).
27. A. Mussot, E. Lantz, H. Maillotte, T. Sylvestre, C. Finot, and S. Pitois, *Opt. Express* **12**, 2838 (2004).
28. K. Tai, A. Hasegawa, and A. Tomita, *Phys. Rev. Lett.* **56**, 135 (1986).
29. E. M. Dianov, A. Y. Karasik, P. V. Mamyshev, A. M. Prokhorov, V. N. Serkin M. F. Stelmakh, and A. A. Fomichev, *JETP Lett.* **41**, 294 (1985).
30. N. Korneev, E. A. Kuzin, B. Ibarra-Escamilla, and M. B. Flores-Rosas, *Opt. Express* **16**, 2636 (2008).
31. M. H. Frosz, O. Bang, and A. Bjarklev, *Opt. Express* **14**, 9391 (2006).
32. F. Biancalana, D. V. Skryabin, and A. V. Yulin, *Phys. Rev. E* **70**, 016615 (2004).
33. B. A. Cumberland, J. C. Travers, S. V. Popov, and J. R. Taylor, *Opt. Express* **16**, 5954 (2008).
34. A. Kudlinski, A. K. George, J. C. Knight, J. C. Travers, A. B. Rulkov, S. V. Popov, and J. R. Taylor, *Opt. Express* **14**, 5715 (2006).
35. A. Kudlinski and A. Mussot, *Opt. Lett.* **33**, 2407 (2008).

36. A. V. Gorbach, D. V. Skryabin, J. M. Stone, and J. C. Knight, *Opt. Express* **14**, 9854 (2006).
37. A. Ortigosa-Blanch, J. C. Knight, and P. St. J. Russel, *J. Opt. Soc. Am. B* **19**, 2567 (2002).
38. G. Geny, M. Lehtonen, H. Ludvigsen, J. Broeng, and M. Kaivola, *Opt. Express* **10**, 1083 (2002).
39. A. V. Gorbach, D. V. Skyabin, J. M. Stone, and J. C Knight, *Opt. Express* 14, 9854 (2006).
40. Y. Kodama and A. Hasegawa, *IEEE J. Quant. Elect.* **23**, 510 (1987).
41. A. Husakou and J. Herrmann, *Phys. Rev. Lett.* **87**, 203901 (2001).
42. M. H. Frosz, P. Falk, and O. Bang, *Opt. Express* **13**, 6181 (2005).
43. G. Genty, M. Lehtonen, and H. Ludvigsen, *Opt. Lett.* **30**, 756 (2005).
44. A. C. Judge, O. Bang, and C. M. de Sterke, *J. Opt. Soc. Am. B* **27**, 2195 (2010).
45. S. Roy, S. K. Bhadra, K. Saitoh, M. Koshiba, and G. P. Agrawal, *Opt. Express* **19**, 10443 (2011).
46. D. V. Skryabin and A. V. Yulin, *Phys. Rev. E* **72**, 016619 (2005).
47. J. H. V. Price, T. M. Monro, H. Ebendorff-Heidepriem, F. Poletti, P. Horak, V. Finazzi, J. Y. Y. Leong, P. Petropoulos, J. C. Flanagan, G. Brambilla, X. Feng, and D. J. Richardson, *IEEE J. Sel. Top. Quantum Electron.* **23**, 738 (2007).
48. J. M. Dudley and S. Coen, *J. Sel. Top. Quantum Electron.* **8**, 651 (2002).
49. A. L. Gaeta, *Opt. Lett.* **27**, 924 (2002).
50. A. V. Husakou and J. Herrmann, *J. Opt. Soc. Am. B* **19**, 2171 (2002).
51. J. Hult, *J. Lightwave Technol.* **25**, 3770 (2007).
52. R. H. Stolen, J. P. Gordon, W. J. Tomlinson, and H. A. Haus, *J. Opt. Soc. Am. B* **6**, 1159 (1989).
53. G. P. Agrawal, *Nonlinear Fiber Optics*, 4th ed., Academic Press, San Diego (2007).
54. K. J. Blow and D. Wood, *IEEE J. Quantum Electron.* **25**, 2665 (1989).
55. Q. Lin and G. P. Agrawal, *Opt. Lett.* **31**, 3086 (2006).
56. U. Moller et al. *Opt. Express* **23**, 3282 (2015).
57. P. Domachuk et al. *Opt. Express* **16**, 7161 (2008).
58. C. Xia et al. *Opt. Lett.* **31**, 2553 (2006).
59. H. Saghaei, M. Kazem M. Farshi, M. Heidari, and M. N. Moghadasi, *IEEE J. Sel. Topics Quantum Electron.* **22**, 4900508 (2016).
60. S. Rodrigues, M. Facão and M. Ferreira, *J. Nonlinear Opt. Phys. Mater.* **26**, 1750049 (2017).
61. J. M. Dudley and S. Coen, *Opt. Lett.* **27**, 1180 (2002).
62. J. M. Dudley, G. Genty, and S. Cohen, *Rev. Modern Phys.* **78**, 1135 (2006).
63. B. Barviau, B. Kibler, and A. Picozzi, *Phys. Rev. A* **79**, 063840 (2009).
64. C. Michel, B. Kibler, and A. Picozzi, *Phys. Rev. A* **83**, 023806 (2011).
65. B. Kibler, C. Michel, A. Kudlinski, B. Barviau, G. Millot, and A. Picozzi, *Phys. Rev. E* **84**, 066605 (2011).
66. A. Picozzi, S. Pitois, and G. Millot, *Phys. Rev. Lett.* **101**, 093901 (2008).
67. B. Barviau, B. Kibler, A. Kudlinski, A. Mussot, G. Millot, and A. Picozzi, *Opt. Express* **17**, 7392 (2009).
68. M. H. Reeve, A. R. Hunwicks, W. Zhao, S. G. Methley, L. Bickers, and S. Hornung, *Electron. Lett.* **24**, 389 (1988).
69. K. H. Han, E. S. Son, H. Y. Choi, K. W. Lin, and Y. C. Chung, *IEEE Photon. Technol. Lett.* **16**, 2380 (2004).
70. S. S. Wagner and T. Chapuran, *Electron. Lett.* **26**, 696 (1990).
71. S. Kaneko, J. Kani, K. Iwatsuki, A. Ohki, M. Sugo, and S. Kamei, *J. Lightwave Technol.* **24**, 1295 (2006).
72. J. S. Lee, Y. C. Chung, and D. J. DiGiovanni, *IEEE Photon. Technol. Lett.* **5**, 1458 (1993).

73. H. Sanjoh, H. Yasaka, Y. Sakai, K. Sato, H. Ishii, and Y. Yoshikuni, *IEEE Photon. Technol. Lett.* **9**, 818 (1997).
74. L. Boivin, M. Wegmuller, M. C. Nuss, and W. H. Knox, *IEEE Photon. Technol Lett.* **11**, 466 (1999).
75. K. Tamura, E. Yoshida, and M. Nakazawa, *Electron. Lett.* **32**, 1691 (1996).
76. J. J. Veselka and S. K. Korotky, *IEEE Photon. Technol. Lett.* **10**, 958 (1998).
77. S. Kawanishi, H. Takara, K. Uchiyama, I. Shake, and M. Mori, *Electron. Lett.* **35**, 826 (1999).
78. L. Boivin, S. Taccheo, C. R. Doerr, P. Schiffer, L. W. Stulz, R. Monnard, and W. Lin, *Electron. Lett.* **36**, 335 (2000).
79. T. Morioka, K. Uchiyama, S. Kawanishi, S. Suzuki, M. Saruwatari, *Electron. Lett.* **31**, 1064 (1995).
80. S. Kawanishi, H. Takara, K. Uchiyama, I. Shake, O. Kamatani, and H. Takahashi, *Electron. Lett.* **33**, 1716 (1977).
81. K. Takada, M. Abe, T. Shibata, and T. Okamoto, *Electron. Lett.* **38**, 572 (2002).
82. K. Mori, K. Sato, H. Takara, and T. Ohara, *Electron. Lett.* **39**, 544 (2003).
83. R. R. Alfano (Ed.), *The Super Continuum Laser Sources*; Springer-Verlag, Berlin (1989).
84. H. Takara, T. Ohara, K. Mori, K. Sato, E. Yamada, K. Jinguji, Y. Inoue, T. Shibata, T. Morioka, and K.-I. Sato, *Electron. Lett.* **36**, 2089 (2000).
85. S. Sotobayashi, W. Chujo, A. Konishi, T. Ozeki, *J. Opt. Soc. Am. B* **19**, 2803 (2002).

6 Optical Pulse Amplification

6.1 INTRODUCTION

In an early stage, the ultimate capacity limits of optic-optic communication systems were determined by the spectral bandwidth of the signal source and of the fundamental fiber parameters: loss and dispersion. However, in the mid-1980s of the last century, the international development had reached a state at which not only dispersion-shifted fibers but also spectrally pure signal sources were available. In such circumstances, the remaining limitations for long-haul lightwave systems by that time were imposed by fiber loss. These limitations have traditionally been overcome by periodic regeneration of the optical signals at repeaters applying conversion to an intermediate electric signal. However, because of the complexity and high cost of such regenerators, the need for optical amplifiers soon became obvious.

Several means of obtaining optical amplification had been suggested since the 1970s, including direct use of the transmission fiber as gain medium through nonlinear effects, semiconductor amplifiers, or doping optical waveguides with an active material (rare-earth ions) that could provide gain. In this chapter, we provide an overview of the main features of fiber optical amplifiers based on stimulated Raman scattering (SRS), stimulated Brillouin scattering (SBS), and four-wave mixing (FWM) effects.

6.2 FIBER RAMAN AMPLIFIERS

Raman amplification in optical fibers was demonstrated early in the 1970s by Stolen and Ippen [1]. The benefits from Raman amplification were elucidated by many research papers in the mid-1980s [2–8]. However, due principally to the its poor pumping efficiency and the scarcity of high-power pumps at appropriate wavelengths, much of that work was overtaken by erbium-doped fiber amplifiers (EDFAs) by the late 1980s. Then, in the mid-1990s, the development of suitable high-power pumps and the availability of higher Raman gain fibers motivated a renewed interest on Raman amplification. Actually, high Raman gain has been observed in several materials and structures, including silica [9,10], fluoride [11,12], tellurite [13–16], chalcogenide [17,18] and photonic crystal fibers [19,20], and have put it into practice [21–25].

Figure 6.1 shows schematically a fiber Raman amplifier. The pump and the signal beams are injected into the fiber through a wavelength division multiplexing (WDM) fiber coupler. The case illustrated in Figure 6.1 shows the two beams co-propagating inside the fiber, but the counterpropagating configuration is also possible. In fact, the SRS can occur in both directions, forward and backward. It was confirmed experimentally that the Raman gain is almost the same in the two cases [26].

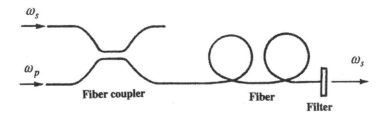

FIGURE 6.1 Schematic of a fiber Raman amplifier.

As seen in Chapter 2, the Raman gain in silica fibers extends over a frequency range of about 40 THz. Optical signals whose bandwidths are of this order or less can be amplified using the Raman effect if a pump wave with the right wavelength is available. The Raman gain depends on the relative state of polarization of the pump and signal fields. The peak value of the Raman gain decreases with increasing the pump wavelength, and it is about 6×10^{-14} m/W in the wavelength region around 1.5 μm.

Equations (2.50) and (2.51) describing the interaction between the pump and signal waves can be rewritten in terms of the involved optical powers in the form:

$$\frac{dP_s}{dz} = -\alpha_s P_s + \frac{g_R}{A_{eff}} P_p P_s \qquad (6.1)$$

$$\frac{dP_p}{dz} = -\alpha_p P_p - \frac{\omega_p}{\omega_s} \frac{g_R}{A_{eff}} P_s P_p \qquad (6.2)$$

where A_{eff} is the effective cross-sectional area of the fiber mode, α_s and α_p are absorption coefficients that account for the fiber loss at the signal and pump frequencies, respectively, and the signal wave is considered to be co-propagating with the pump wave. Assuming that $\alpha_s = \alpha_p = \alpha$ (which is a reasonable approximation around the 1.5 μm wavelength region in a low-loss fiber), the following approximate solutions of Eqs. (6.1) and (6.2) can be written when $P_p(0) \gg P_s(0)$ [27]:

$$P_p(z) = \frac{P_p(0)\exp(-\alpha z)}{1 + F(z)} \qquad (6.3)$$

$$P_s(z) = \frac{P_s(0)\exp\left(\Gamma(1 - e^{-\alpha z}) - \alpha z\right)}{1 + F(z)} \qquad (6.4)$$

where

$$\Gamma = \frac{g_R P_p(0)}{\alpha A_{eff}} \qquad (6.5)$$

Optical Pulse Amplification

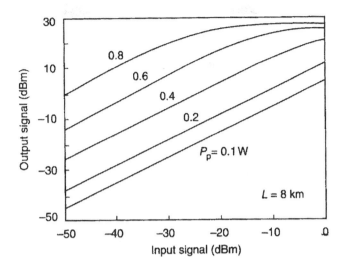

FIGURE 6.2 Transfer characteristics of a Raman amplifier with length $L = 8$ km for several values of the input pump power. The amplifier parameters are: $g_R = 6.7 \times 10^{-14}$ m/W, $\alpha = 0.2$ dB Km^{-1}, $A_{eff} = 30 \times 10^{-12}$ m^2, $\lambda_p = 1.46$ μm, and $\lambda_s = 1.55$ μm.

and

$$F(z) = \frac{\omega_p P_s(0)}{\omega_s P_p(0)} \exp\left(\Gamma(1 - e^{-\alpha z})\right) \tag{6.6}$$

The dimensionless parameter $F(z)$ accounts for the effects of pump depletion and the attendant saturation of the gain seen by the signal.

Figure 6.2 shows the output signal power $P_s(L)$ versus the input signal power $P_s(0)$ for a Raman amplifier with length $L = 8$ km and several values of the input pump power. Typical values $g_R = 6.7 \times 10^{-14}$ m/W, $\alpha = 0.2$ dB km^{-1}, $A_{eff} = 30 \times 10^{-12}$ m^2, $\lambda_p = 1.46$ μm, and $\lambda_s = 1.55$ μm were assumed. The transfer characteristics are linear for $P_p(0) = 0.1, 0.2,$ and 0.4 W, but the effects of pump depletion become discernible for input signal levels $P_s(0) > -15$ dBm when $P_p(0) = 0.6$ W. For a pump power $P_p(0) = 0.8$ W the linear behavior of the Raman amplifier cannot be observed even for signal power levels as low as -25 dBm.

Since in the absence of Raman amplification the signal power at the amplifier output would be $P_s(L) = P_s(0)\exp(-\alpha L)$, the amplifier gain is given by

$$G_R = \frac{P_s(L)}{P_s(0)\exp(-\alpha L)} = \frac{\exp\left(g_R P_p(0) L_{eff} / A_{eff}\right)}{1 + F(L)} \tag{6.7}$$

where L_{eff} is the effective interaction length, given by Eq. (2.54). The amplifier gain is seen to be a function of the input signal power (i.e., a saturation nonlinearity) through the term $F(L)$.

When $F(L) \gg 1$ we have from Eq. (6.4) that

$$P_s(L) = \frac{\omega_s}{\omega_p} P_p(0) \exp(-\alpha L) \tag{6.8}$$

In this case, the output signal reaches the pump level irrespective of the input signal level. This implies that any spontaneous Raman scatter in the fiber will be amplified up to power levels comparable to that of the pump, which must be avoided in practice.

On the other hand, when $F(L) \ll 1$ we have from Eq. (6.4) that

$$P_s(L) = P_s(0)\exp(g_R L_{eff} P_p(0)/A_{eff} - \alpha L) \tag{6.9}$$

In this case, the amplifier gain is

$$G_R = \frac{P_s(L)}{P_s(0)\exp(-\alpha L)} = \exp(g_R L_{eff} P_0 / A_{eff}) \tag{6.10}$$

From Eq. (6.10), the Raman gain in decibels is expected to increase linearly with the pump input power, which was confirmed experimentally [28].

Considering its broad bandwidth (>5 THz), fiber Raman amplifiers can be used to amplify several channels simultaneously in a multichannel communication system. The same characteristic also makes such amplifiers suitable for amplification of short optical pulses. Moreover, Raman amplifiers can in principle have a fully configurable gain profile determined by the Raman gain spectra produced by multiple pump lasers. In practice, CW power at several wavelengths is provided by a set of high-power semiconductor lasers located at the pump stations. The wavelengths of pump lasers should be in the vicinity of 1450 nm for amplifying optical signals in the 1550 nm spectral region. These wavelengths and pump-power levels are chosen to provide a uniform gain over the entire C band. Raman amplifiers with gain bandwidths greater than 100 nm were demonstrated [29,30].

When using a broad pump spectrum an important issue is the interaction between the pumps, which affects the noise properties of the amplifier. Problems arise particularly due to FWM between the pumps, since it can create light at new frequencies within the signal band. This new light can interfere with the signal channels, producing beat noise [31].

Another problem can arise from SBS. Since the intrinsic gain coefficient for stimulated SBS is two orders of magnitude larger than that for SRS. In some cases, the SBS may occur at lower pump powers and may significantly affect the SRS process [32]. In particular, the Raman gain may not only be reduced but also become unstable. However, the SBS can be suppressed if a pump laser with linewidth broader than the Brillouin bandwidth is employed.

The dominant noise light in fiber Raman amplification is due to the amplified spontaneously Raman scattered light. In fact, a part of the pump energy is spontaneously converted into Stokes radiation extending over the entire bandwidth of the Raman gain spectrum and is amplified together with the signal. The output thus consists not only of the desired signal but also of background noise extending over a wide frequency range (~10 THz or more).

It has been shown that, for both forward and backward pumpings, the noise light power is equivalent to a hypothetical injection of a single-photon per unit frequency at the fiber input end for forward pumping, and at some distance away from the fiber output for backward pumping. Assuming that $G_R \gg 1$ and $\alpha L \ll 1$, the noise light powers for forward pumping, $P_{fasp}(L)$, and backward pumping, $P_{basp}(L)$, are approximately given by [33,34]

$$P_{fasp}(L) \approx h v_s \Delta v_R (G_R - 1) \exp(-\alpha L) \qquad (6.11)$$

$$P_{basp}(L) \approx h v_s \Delta v_R (G_R - 1) / \ln G_R \qquad (6.12)$$

where Δv_R is the Raman gain bandwidth, h is Planck's constant, and v_s is the signal frequency. We can observe from Eqs. (6.11) and (6.12) that the noise light power in the case of forward pumping decreases as the fiber length is increased. However, in the case of backward pumping, it depends mainly on the Raman gain, being nearly independent of fiber length and loss.

Raman amplifiers can be realized considering two main options. One is the lumped Raman amplifier, in which all the pump power is confined to a relatively short fiber element that is inserted into the transmission line to provide gain. The primary use of the lumped amplifier is to open new wavelength bands between about 1280 and 1530 nm, a wavelength range that is inaccessible by EDFAs. The other option is the distributed Raman amplifier, which utilizes the transmission fiber itself as the Raman gain medium to obtain amplification. This option has the merit of reducing the overall excursion experienced by the signal power. Consequently, nonlinear effects are reduced at higher signal levels, whereas the signal-to-noise ratio (SNR) remains relatively high at lower signal levels.

Short fiber Raman amplifiers can be realized using some of the highly nonlinear fibers (HNLFs) developed during the recent years. The required pump power is also lower in such cases. In particular, standard fibers based on tellurite or arsenic selenide (As_2Se_3) glass compositions have been drawn to obtain the Raman amplification and lasing characteristics [13,35–38]. The Raman gain efficiency is further increased using microstructured fibers. A peak gain of 10 dB was reported recently for a 1.1-m-long As_2Se_3 microstructure optical fiber pumped at 1500 nm with an input power of 500 mW [38].

In recent years, much work has been carried out in order to extend the operating wavelength of fiber lasers and amplifiers toward the longer mid-infrared wavelength region [39–41], driven by a large number of promising applications, including LIDAR, gas sensing, and optical communication. Thulium-doped fiber lasers and amplifiers operate efficiently between 1.8 and 2.1 µm wavelength regions [42–45], whereas holmium-doped fiber lasers and amplifiers have been demonstrated to be efficient at >2.1 µm wavelength [46–48]. Raman fiber lasers or amplifiers constitute a promising alternative for operation at >2.1 µm wavelength. Actually, the SRS has permitted the development of a wide variety of fiber lasers and amplifiers at wavelengths for which there are no rare-earth-doped gain media available [49–51]. The principle is to wavelength convert the output of a rare-earth-doped fiber laser or amplifier to the required output wavelength by using the first order or cascaded Raman Stokes shifts [52]. Cascaded Raman wavelength shifting up to the fourth order ranging from 2.1 to 2.4 μm in a low-loss chalcogenide suspended-core fiber has been demonstrated [53]. In a 2014 experiment, a 2147-nm silica-based Raman all-fiber amplifier with output power of 14.3 W directly pumped with a 1963 nm continuous wave (CW) thulium-doped all-fiber master-oscillator power-amplifier was demonstrated. The conversion efficiency was 38.5% from 1963 nm to 2147 nm in such Raman fiber amplifier [10].

In recent years, few-mode distributed Raman amplifiers (FM-DRAs) have attracted an increasing attention [54–57], since they are an attractive amplification solution for long-haul mode-division multiplexing transmission based on few-mode fibers [58–60]. Actually, compared with few-mode erbium-doped fiber amplifiers (FM-EDFAs), FM-DRAs have some fundamental advantages, such as simple configuration, broad gain bandwidth, flexible operation window, and low noise figure (NF) [61].

High-order DRAs use additional pumps at wavelengths with one or several times of Stokes shifts below the conventional first-order pump [31] in order to amplify the first-order pump and to provide gain farther from the pump input. This scheme leads to an improved NF performance [31,62]. In a 2017 experiment, a second-order FM-DRA has been demonstrated [63]. The 1455 and 1360 nm pumps were both backward launched into the a 70-km long 4-LP mode fiber in the forms of two degenerate LP11 modes. Compared with the conventional first-order pumping scheme, the NFs at 1550 nm for LP01 and LP11 modes were improved by 1.2 dB and 1.1 dB, respectively.

6.3 FIBER BRILLOUIN AMPLIFIERS

SBS in optical fibers is a highly efficient nonlinear amplification mechanism with which large gains that can be achieved using pump powers of only a few milliwatts. Such nonlinear process can be used to construct a fiber Brillouin amplifier (FBA), which is basically an optical fiber in which the pump and signal waves propagate in opposite directions [64–66]. The only required condition is that the spacing between the pump and signal frequencies must correspond to the Brillouin shift in that fiber. FBAS find applications in microwave photonics [67], radio-over-fiber technology [68,69], and narrowband optical filtering [70,71]. In addition, FBAs proved to be useful for the generation of millimeter-wave signals [72,73], as well as for the implementation of tunable slow-light delay buffers [74].

In the SBS process, the power evolution of the pump and signal waves is governed by the following coupled equations, which can be obtained from Eqs. (2.55) and (2.56) [27]:

$$\frac{dP_S}{dz} = -\frac{g_B(\Delta\omega_S)}{A_{eff}} P_p P_S + \alpha P_S \quad (6.13)$$

$$\frac{dP_p}{dz} = -\frac{g_B(\Delta\omega_S)}{A_{eff}} P_S P_p - \alpha P_p \quad (6.14)$$

where $g_B(\Delta\omega_S)$ is the Brillouin gain coefficient, given by:

$$g_B(\Delta\omega_S) = \frac{(\Gamma_B/2)^2}{(\Delta\omega_S)^2 + (\Gamma_B/2)^2} g_{B0} \quad (6.15)$$

As shown by Eq. (6.15), the spectrum of the Brillouin gain is Lorentzian with a full width half maximum $\Delta v_B = \Gamma_B/2\pi$ determined by the acoustic phonon lifetime. For bulk silica, the width is expected to be about 17 MHz at 1.5 µm. However, several experiments have shown much larger bandwidths for silica-based fibers, which in some cases can exceed 100 MHz [75,76]. Considering the parameter values typical of fused silica, the Brillouin coefficient g_{B0} is estimated to be about 2.5×10^{-11} m/W, which is between two and three orders of magnitude larger than the Raman gain coefficient in silica fibers at $\lambda_p = 1.55$ µm. Larger values of the Brillouin gain coefficient can yet be obtained in some non-silica-based fibers. For example, a peak value of 1.6989×10^{-10} m/W was reported for a tellurite fiber [77], whereas a value of $\sim 6.08 \times 10^{-9}$ m/W, which is more than 200 times larger than that of silica, has been measured in a single-mode As_2Se_3 chalcogenide fiber [78].

The exact solution of Eqs. (6.13) and (6.14) is known only for lossless media [79–81], which is not the case for real FBAs. However, considering that $\alpha \ll 1$, the following approximate solutions can be derived [27,64]:

$$P_S(z) = P_S(0)D(z)e^{\alpha z} \quad (6.16)$$

$$P_p(z) = P_p(0)D(z)H(z)e^{-\alpha z} \quad (6.17)$$

where

$$D(z) = \frac{[P_p(0) - P_S(0)]}{[P_p(0)H(z) - P_S(0)]} \quad (6.18)$$

and

$$H(z) = \exp\{g_B[P_p(0) - P_S(0)][1 - \exp(-\alpha z)]/(A_{\it eff}\alpha)\} \quad (6.19)$$

The Brillouin amplifier gain is given from Eq. (6.16) as

$$G_B = \frac{P_S(0)}{p_S(L)e^{-\alpha L}} = \frac{1}{D(L)} \quad (6.20)$$

where $D(L)$ is given by Eq. (6.18) with $z = L$.

Figure 6.3 illustrates the dependence of the Brillouin amplifier gain on the input pump power for several amounts of the signal detuning from the gain peak and for two values of the signal power. A fiber length $L = 5$ km, a gain bandwidth $\Delta v_B = 30$ MHz, typical parameter values $g_B(0) = 2.5 \times 10^{-11}$ m/W, $\alpha = 0.2$ dB km^{-1}, and $A_{\it eff} = 30 \times 10^{-12}$ m^2 were assumed. We observe that the saturated gain shows a significant dependence on the input signal power. An input signal of 1 nW, which

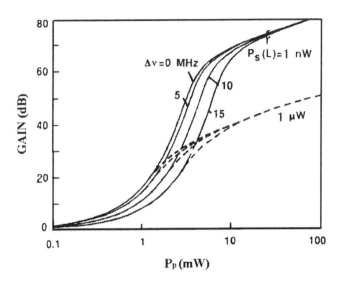

FIGURE 6.3 Brillouin amplifier gain against the input pump power for an amplifier length $L = 5$ km and different amounts of detuning from the gain center. The full curves and the dashed curves correspond to input signal powers $P_s(L) = 1$ nW and $P_S(L) = 1$ μW, respectively.

is comparable to the original value of spontaneous emission, can be amplified by about 60 dB for a pump power $P_p(0) \approx 5$ mW. However, the amplifier gain can be considerably reduced when the signal is detuned from the gain peak. For example, for an input signal power $P_s(L) = 1$ nW and a pump power $P_p(0) \approx 3$ mW a detuning of 15 MHz (which corresponds to a reduction of the gain coefficient to half of its maximum value) determines a reduction of about 25 dB on the amplifier gain.

The narrow Brillouin linewidth strictly limits the bandwidth of data signals that can be amplified in an FBA. However, as previously mentioned, the intrinsic Brillouin linewidth is generally enhanced by compositional inhomogeneities in the fiber and can be intentionally extended by more than one order of magnitude by applying frequency modulation to the pump laser. Of course, the bandwidth enhancement is accompanied by a reduction of the peak gain. Consequently, a higher pump power will be necessary to achieve the same gain.

In a 1986 experiment, 5 dB of net gain was obtained in a 37.5-km-long FBA using a pump power of 3.5 mW [64]. The gain occurred over a 150 MHz bandwidth, due to fiber nonuniformity. Later, the technique of broadening the gain bandwidth by pump laser frequency sweeping was demonstrated and applied to the amplification of data-carrying signals at 10 and 90 Mb/s [65]. The receiver sensitivities showed the full improvement of the gain achieved in these measurements [82], since the signals were attenuated before reaching the receiver.

Brillouin amplifiers made with conventional silica fibers use generally fiber lengths of some kilometers. However, shorter Brillouin amplifiers can be realized using fibers made of highly nonlinear glasses [27,83–85]. Tellurite fibers offer the advantage of having a relatively low background loss of 0.02 dB/m at 1550 nm [13,36], which means a larger effective length and consequently a greater amplifier gain.

Optical Pulse Amplification

A Brillouin gain of 29 dB and a linewidth of 20.98 MHz were recently reported using a tellurite fiber with a length of only 100 m and a pump power of 10 mW at 1550 nm [79].

In order to describe the spontaneous emission noise, we must add a spontaneous emission term to the equations describing the evolution of the Stokes and pump waves. The amplified spontaneous scattered power per unit frequency is then obtained in the form [86]

$$P_{asp}(v,z) = (hv/A_e)g_B(\delta)(N+1)G(v,z)\int_z^L P_p(z')[G(v,z')]^{-1} dz' \quad (6.21)$$

where

$$G(v,z) = \exp\left\{\int_z^L \left[P_p(z)g_B(\delta)/A_e - \alpha\right]dz\right\} \quad (6.22)$$

is the gain function, h is Planck's constant, and N is the thermal equilibrium number of acoustic phonons, given by [86]

$$N = \frac{1}{\exp(hv_B/kT)-1} \approx kT/hv_B \approx 500 \quad (6.23)$$

where k is the Boltzmann constant, and T is the absolute temperature. The total amplified signal and spontaneous powers are given by integration of $P_s(v,z)$ over the gain profile.

Similarly to the behavior of the gain, the noise power depends significantly on the signal magnitude in the saturation regime and becomes particularly high for low signal powers. This high noise level, which is about 20 dB above that of an ideal amplifier, imposes some limitations on the use of FBA as a receiver preamplifier.

Fiber Brillouin amplifiers can provide high gains at low pump powers. However, the gain bandwidth is small, and the amplified spontaneous emission noise is substantially larger than that observed with other amplifiers, leading to a NF quite large (>15 dB). This fact limits their usefulness as a preamplifier. Use as in-line amplifiers at high signal levels is possible in some configurations. Some problems in this application concerns with the narrow bandwidth and the small saturation power of the FBAs. In particular, the Brillouin linewidth strictly limits the bandwidth of data signals that can be amplified in an FBA.

The Brillouin gain bandwidth can be intentionally extended by more than one order of magnitude by applying frequency modulation to the pump laser. Of course, the bandwidth enhancement is accompanied by a reduction of the peak gain. Consequently, a higher pump power will be necessary to achieve the same gain.

The narrow bandwidth of the FBAs can be used to advantage incoherent lightwave systems to amplify selectively the carrier of a phase- or amplitude-modulated wave, thus allowing homodyne detection, using the amplified carrier as the local oscillator [87].

The scheme should work well at bit rates >100 Mb/s because the modulation sidebands then fall outside the amplifier bandwidth, and the optical carrier can be amplified selectively. However, an inevitable accompaniment of this narrow-band amplification is the nonlinear phase shift induced by the pump on the signal [64]. This phase shift imposes the most stringent limit on pump- and signal-frequency stability when FBAs are used for self-homodyne detection. In the case of amplitude-shift keyed signals, for example, a phase stability of about 0.1 rad for the amplified signal carrier may be required, which corresponds to a pump-signal frequency offset <100 kHz. The nonlinear phase shift can be used with advantage in some self-homodyne schemes, which require some specific adjustment of the carrier phase. In the case of a self-homodyne coherent receiver for phase-shift keyed signals, for example, a quadrature phase correction is required [88]. Typically, this requires a detuning of only a few MHz from the SBS gain peak, well within the Brillouin gain bandwidth.

Optical filtering with a narrow band of about MHz to several GHz is highly demanded for a range of important applications, especially in optical fiber sensing [89,90]. Because the SBS effect has an inherent narrow bandwidth on the order of 10 MHz, it is a perfect choice for extensive applications, such as high-resolution spectroscopy and microwave photonics [91–93].

Another application of the narrow line width associated with the Brillouin gain is as a tunable narrow-band filter for channel selection in a densely packed multichannel communication system [70,71,94]. A channel can be selectively amplified through Brillouin amplification by launching a pump beam at the receiver end so that it propagates inside the fiber in a direction opposite to that of the multi-channel signal. Different channels can be selectively amplified by tuning the pump laser. The adjustable bandwidth and high out-of-band rejection can be used to advantage in this case. In another approach, the fiber Brillouin amplifier can be used to selectively amplify multiple subcarrier-multiplexed channels carried by a single optical carrier within a WDM signal, or to amplify a backward-propagating supervisory signal.

FBAs can be used also to realize slow light. Some important applications for slow light have been proposed, including optical buffering, data synchronization, optical memories, and signal processing [74,95–99]. Compared with other approaches, fiber-based SBS slow light offers several advantages: the use of an optical fiber provides long interaction lengths, the involved resonance can be created at any wavelength by changing the pump wavelength, and the process can be achieved at room temperature [98–100]. The use of fibers made of highly nonlinear glasses is a good option to enhance the slow-light generation. The use of a 2-m bismuth-oxide HNLF was reported to generate 29-dB Brillouin gain and the resultant optical delay of 46 ns with 410-mW pump power [101]. In another experiment, a Brillouin gain of 43 dB was achieved with only 60-mW pump power in a 5-m-long As_2O_3 chalcogenide fiber, which leads to an optical time delay of 37 ns [102].

6.4 FIBER PARAMETRIC AMPLIFIERS

FWM can be used with advantage to realize fiber-optical parametric amplifiers (FOPAs). A FOPA offers a wide gain bandwidth and may be tailored to operate at any wavelength [103–107]. For example, recent FOPA demonstrations have highlighted

Optical Pulse Amplification

>100 nm gain bandwidth between 1555 and 1665 nm using a single pump [108], phase sensitive operation with NF below the 3 dB quantum limit [109], and the use of phase-conjugated idlers produced via the gain process for transmission nonlinearity compensation [110,111].

Using the result given by Eq. (2.41) for the pump fields and the transformation

$$V_j = U_j \exp\{-2i\gamma(P_1+P_2)z\}, (j=3,4), \qquad (6.24)$$

we obtain from Eqs. (2.38) and (2.39) the following coupled equations governing the evolution of the signal and idler waves:

$$\frac{dV_3}{dz} = 2i\gamma\sqrt{P_1P_2}\exp(-i\kappa z)V_4^* \qquad (6.25)$$

$$\frac{dV_4^*}{dz} = -2i\gamma\sqrt{P_1P_2}\exp(i\kappa z)V_3 \qquad (6.26)$$

where the effective phase mismatch κ is given by Eq. (2.44). If only the signal and the pumps are launched at $z=0$, that is, assuming that $V_4^*(0)=0$, Eqs. (6.36) and (6.37) have the following solutions:

$$V_3(z) = V_3(0)\bigl(\cosh(gz)+(i\kappa/2g)\sinh(gz)\bigr)\exp(-j\kappa z/2) \qquad (6.27)$$

$$V_4^*(z) = -i(\gamma/g)\sqrt{P_1P_2}V_3(0)\sinh(gz)\exp(j\kappa z/2) \qquad (6.28)$$

where g is the parametric gain, given by

$$g = \sqrt{4\gamma^2 P_1 P_2 - \left(\frac{\kappa}{2}\right)^2} \qquad (6.29)$$

The maximum gain occurs for perfect phase matching ($\kappa=0$) and is given by

$$g_{max} = 2\gamma\sqrt{P_1P_2} \qquad (6.30)$$

From Eqs. (6.27) we can write the following result for the signal power, $P_3 = |V_3|^2$:

$$P_3(z) = P_3(0)\left[1+\left(1+\frac{\kappa^2}{4g^2}\right)\sinh^2(gz)\right] = P_3(0)\left[1+\frac{4\gamma^2 P_1 P_2}{g^2}\sinh^2(gz)\right] \qquad (6.31)$$

The idler power $P_4 = |V_4|^2$ can be obtained by noting from Eqs. (6.27) and (6.28) that $P_3(z) - P_4(z) = \text{constant} = P_3(0)$. The result is:

$$P_4(z) = P_3(0) \frac{4\gamma^2 P_1 P_2}{g^2} \sinh^2(gz) \tag{6.32}$$

The unsaturated single-pass gain of a FOPA of length L becomes,

$$G_P = \frac{P_3(L)}{P_3(0)} = 1 + \frac{4\gamma^2 P_1 P_2}{g^2} \sinh^2(gL) \tag{6.33}$$

According to Eq. (6.29), amplification ($g > 0$) occurs only in the case

$$|\kappa| < 4\gamma \sqrt{P_1 P_2} \tag{6.34}$$

Since $|\kappa|$ must be small, this also means good phase matching. For $|\kappa| > 4\gamma\sqrt{P_1 P_2}$, the parametric gain becomes imaginary, and there is no longer amplification but rather a periodical power variation of the signal and idler waves.

In the case of a single pump, the effective phase mismatch is given by Eq. (2.44), and the parametric gain coefficient becomes:

$$g^2 = \left[(\gamma P_p)^2 - (\kappa/2)^2\right] = -\Delta k \left(\frac{\Delta k}{4} + \gamma P_p\right) \tag{6.35}$$

where P_p is the pump power. The maximum value of the parametric gain occurs when $\Delta k = -2\gamma P_p$ and is given by

$$g_{max} = \gamma P_p \tag{6.36}$$

The unsaturated single-pass gain may be written as

$$G_p = \frac{P_3(L)}{P_3(0)} = 1 + \left[\frac{\gamma P_p}{g} \sinh(gL)\right]^2 \tag{6.37}$$

In the case of perfect phase matching ($\kappa = 0$) and assuming that $\gamma P_p L \gg 1$, the unsaturated single-pass gain becomes

$$G_p \approx \sinh^2(gL) \approx \frac{1}{4} \exp(2\gamma P_p L) \tag{6.38}$$

Considering the approximation given by Eq. (2.45) for the linear phase mismatch, the FOPA bandwidth $\Delta\Omega$ can be obtained from the maximum effective phase mismatch,

$$\kappa_m = \beta_2 (\Omega_s + \Delta\Omega)^2 + 2\gamma P_p, \tag{6.39}$$

with $\Delta\Omega \ll \Omega_s$ and Ω_s given by Eq. (2.47). This maximum value of the effective phase mismatch occurs when the parametric gain in Eq. (6.35) vanishes, which gives $\kappa_m = 2\gamma P_p$. In these circumstances, the FOPA bandwidth is approximately given by

$$\Delta\Omega \approx \sqrt{\frac{\gamma P_p}{2|\beta_2|}} \qquad (6.40)$$

Equation (6.40) shows that the FOPA bandwidth can be increased by increasing the nonlinear parameter γ and reducing the $|\beta_2|$. This is the reason why modern FOPAs use generally HNLFs, and the pump wavelengths are chosen near the zero-dispersion wavelength of the fiber. Moreover, Eq. (6.37) shows that, for a fixed value of the FOPA gain, the value of $\gamma P_p L$ must be kept constant. In such case, the amplifier bandwidth will increase with decreasing L. Using a HNLF provides the possibility of simultaneously decreasing L and increasing the nonlinear parameter γ. For example, a value $\gamma P_p L = 10$ can be achieved with a pump power of 1 W and a HNLF with $L = 1$ km and $\gamma = 10$ W^{-1}/km. The bandwidth will be increased 20 times using, for instance, a fiber length of 50 m, a pump power of 5 W, and a nonlinear parameter $\gamma = 40$ W^{-1}/km. Such a high value of γ can be achieved using some types of HNLFs.

As the FOPA bandwidth is proportional to the square root of the pump power, it could be, in principle, arbitrarily increased if enough optical power is available. However, such high values of the pump power are limited in practice by SBS. SBS shows a reduced threshold in the case of HNLFs due to the small effective core area of this type of fibers.

The limitations imposed by the SBS process can be partially circumvented if the pump power is distributed between two pumps, instead of being concentrated in one pump. Moreover, a two-pump FOPA offers additional degrees of design freedom, which makes it fundamentally different from the conventional one-pump device [112–117]. Figure 6.4 illustrates the operating principle of a two-pump FOPA, with one pump in the anomalous and the other pump in the normal dispersion regimes. A flat, broadband parametric gain can be generated by controlling three distinct parametric processes. First, the degenerate exchange $2\omega_{1,2} \rightarrow \omega_{1-,2-} + \omega_{1+,2+}$, which is recognized as modulation instability. Second, phase conjugation [115], which allows for symmetric, nondegenerate FWM: $\omega_1 + \omega_2 \rightarrow \omega_{1-} + \omega_{2+}$ and $\omega_1 + \omega_2 \rightarrow \omega_{1+} + \omega_{2-}$. Finally, Bragg scattering [115,118] enables the inherently stable exchanges $\omega_1 + \omega_{2+} \rightarrow \omega_{1+} + \omega_2$ and $\omega_{1-} + \omega_2 \rightarrow \omega_1 + \omega_{2-}$. When the signal frequency is close a pump frequency modulation instability is the dominant parametric process, whereas phase conjugation is responsible for photon creation in spectral regions further from pump frequencies [115].

The polarization sensitivity is an important issue of any fiber communication device. This is indeed a main problem of one-pump FOPA, which presents a high degree of polarization sensitivity. In the case of a two-pump FOPA, the gain is maximized when the pump and signal are copolarized along the entire interaction length. Any deviation from such copolarized state determines a significant reduction of the parametric gain, which can become negligible for a signal polarization orthogonal to two copolarized pumps [119]. A simple and elegant solution

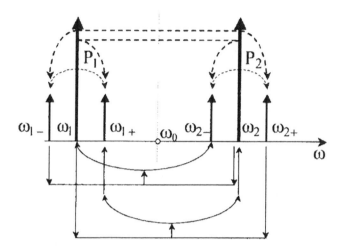

FIGURE 6.4 Two-pump parametric processes. $P_{1,2}$ are pumps, and $1\pm, 2\pm$ are parametric sidebands. Modulation instability is indicated by thin dashed lines, phase conjugation by thick dashed lines, and Bragg scattering by thin solid lines. (Reprinted from *Opt. Fib. Technol.*, 9, Radic, S., and McKinstrie, C.J., 7, Copyright 2003, with permission from Elsevier. Ref. [116].)

to this problem can be achieved using orthogonally multiplexed pumps [119–122]. This configuration results in the near polarization invariance of the parametric process. However, such improvement is achieved at expenses of a significant reduction of the parametric gain, compared with the copolarized two-pump scheme [114]. In a 2017 experiment, polarization-division multiplexed dense wavelength division multiplexing data transmission was demonstrated for the first time in a range of systems incorporating a net-gain polarization-insensitive (PI) FOPA for loss compensation. The PI-FOPA comprised a modified diversity-loop architecture to achieve 15 dB net-gain, and up to 2.3 THz (~18 nm) bandwidth [123,124].

REFERENCES

1. R. H. Stolen and E. P. Ippen, *Appl. Phys. Lett.* **22**, 276 (1973).
2. M. Ikeda, *Opt. Commun.* **39**, 148 (1981).
3. A. R. Chraplyvy, J. Stone, and C. A. Burrus, *Opt. Lett.* **8**, 415 (1983).
4. M. Nakazawa, *Appl. Phys. Lett.* **46**, 628 (1985).
5. M. Nakazawa, T. Nakashima, and S. Seikai, *J. Opt. Soc. Am. B* **2**, 215 (1985).
6. M. L. Dakss and P. Melman, *J. Lightwave Technol.* **3**, 806 (1985).
7. N. A. Olsson and J. Hegarty, *J. Lightwave Technol.* **4**, 391 (1986).
8. Y. Aoki, S. Kishida, and K. Washio, *Appl. Opt.* **25**, 1056 (1986).
9. M. Ikeda, *Opt. Commun.* **39**, 148 (1981).
10. J. Liu, F. Tan, H. Shi, and P. Wang, *Opt. Express* **22**, 28383 (2014).
11. T. L. Cheng, W. Q. Gao, X. J. Xue, T. Suzuki, and Y. Ohishi, *Opt. Fiber Technol.* **36**, 245 (2017).
12. V. Fortin, M. Bernier, D. Faucher, J. Carrier, and R. Vallée, *Opt. Express* **20**, 19412 (2012).

13. A. Mori, H. Masuda, K. Shikano, and M. Shimizu, *J. Lightwave Technol.* **21**, 1300 (2003).
14. G. Zhu, L. Geng, X. Zhu, L. Li, Q. Chen, R. A. Norwood, T. Manzur, and N. Peyghambarian, *Opt. Express* **23**, 7559 (2015).
15. M. Liao, X. Yan, W. Gao, Z. Duan, G. Qin, T. Suzuki, and Y. Ohishi, *Opt. Express* **19**, 15389 (2011).
16. T. L. Cheng, W. Q. Gao, X. J. Xue, T. Suzuki, and Y. Ohishi, *Opt. Mater. Express* **6**, 3438 (2016).
17. S. D. Jackson and G. Anzueto-Sánchez, *Appl. Phys. Lett.* **88**, 221106 (2006).
18. M. Bernier, V. Fortin, M. El-Amraoui, Y. Messaddeq, and R. Vallée, *Opt. Lett.* **39**, 2052 (2014).
19. F. Benabid, J. C. Knight, G. Antonopoulos, and P. St. J. Russell, *Science* **298**, 399 (2002).
20. S. Coen, A. H. L. Chau, R. Leonhardt, J. D. Harvey, J. C. Knight, W. J. Wadsworth, and P. St. J. Russell, *J. Opt. Soc. Am. B* **19**, 753 (2002).
21. G. S. Qin, R. Jose, and Y. Ohishi, *J. Appl. Phys.* **101**, 093109 (2007).
22. R. H. Stolen, C. Lin, and R. K. Jain, *Appl. Phys. Lett.* **30**, 340 (1977).
23. J. D. Shi, X. Feng, P. Horak, K. K. Chen, P. Siong Teh, S. Alam, W. H. Loh, D. J. Richardson, and M. Ibsen, *J. Lightwave Technol.* **29**, 3461 (2011).
24. J. Vieira, R. M. G. M. Trines, E. P. Alves, R. A. Fonseca, J. T. Mendonça, R. Bingham, P. Norreys, and L. O. Silva, *Nat. Commun.* **7**, 10371 (2016).
25. F. E. Robles, K. C. Zhou, M. C. Fischer, and W. S. Warren, *Optica* **4**, 243 (2017).
26. Y. Aoki, S. Kishida, H. Honmou, K. Washio, and M. Sugimoto, *Electron. Lett.* **19**, 620 (1983).
27. M. Ferreira, *Nonlinear Effects in Optical Fibers*; John Wiley & Sons, Hoboken, NJ (2011).
28. A. Tomita, *Opt. Lett.* **8**, 412 (1983).
29. S. Namiki and Y. Emori, *IEEE J. Sel. Top. Quantum Electron.* **7**, 3 (2001).
30. C. Fukai, K. Nakajima, J. Zhou, K. Tajima, K. Kurokawa, and I. Sankawa, *Opt. Lett.* **29**, 545 (2004).
31. J. Bromage, *J. Lightwave Technol.* **22**, 79 (2004).
32. M. F. Ferreira, J. F. Rocha, and J. L. Pinto, *Electron. Lett.* **27**, 1576 (1991).
33. Y. Aoki, *Opt. Quantum Electron.* **21**, S89 (1989).
34. Y. Aoki, *IEEE J. Lightwave Technol.* **6**, 1225 (1988).
35. L. B. Shaw, P. C. Pureza, V. Q. Nghuyen, J. S. Sanghera, and I. D. Aggarwal, *Opt. Lett.* **28**, 1406 (2003).
36. H. Masuda, A. Mori, K. Shikano, and M. Shimizu, *J. Lightwave Technol.* **24**, 504 (2006).
37. P. A. Thielen, L. B. Shaw, J. Sanghers, and I. Aggarwal, *Opt. Express* **11**, 3248 (2003).
38. S. K. Varshney, K. Saito, K. Lizawa, Y. Tsuchida, M. Koshiba, and R. K. Sinha, *Opt. Lett.* **33**, 2431 (2008).
39. S. D. Jackson, *Nat. Photonics* **6**, 423 (2012).
40. C. W. Rudy, A. Marandi, K. L. Vodopyanov, and R. L. Byer, *Opt. Lett.* **38**, 2865 (2013).
41. K. Liu, J. Liu, H. Shi, F. Tan, and P. Wang, *Opt. Express* **22**, 24384 (2014).
42. J. Geng, Q. Wang, T. Luo, S. Jiang, and F. Amzajerdian, *Opt. Lett.* **34**, 3493 (2009).
43. Q. Fang, W. Shi, K. Kieu, E. Petersen, A. Chavez-Pirson, and N. Peyghambarian, *Opt. Express* **20**, 16410 (2012).
44. X. Wang, P. Zhou, X. Wang, H. Xiao, and L. Si, *Opt. Express* **21**, 32386 (2013).
45. J. Liu, J. Xu, K. Liu, F. Tan, and P. Wang, *Opt. Lett.* **38**, 4150 (2013).
46. S. D. Jackson, A. Sabella, A. Hemming, S. Bennetts, and D. G. Lancaster, *Opt. Lett.* **32**, 241 (2007).

47. A. Hemming, N. Simakov, A. Davidson, S. Bennetts, M. Hughes, N. Carmody, P. Davies, L. Corena, D. Stepanov, J. Haub, R. Swain, and A. Carter, CLEO, OSA Technical Digest (online) (Optical Society of America, 2013), paper CW1M.1 (2013).
48. N. Simakov, A. Hemming, W. A. Clarkson, J. Haub, and A. Carter, *Opt. Express* **21**, 28415 (2013).
49. Y. Feng, L. R. Taylor, and D. B. Calia, *Opt. Express* **17**, 23678 (2009).
50. L. Zhang, H. Jiang, S. Cui, and Y. Feng, *Opt. Lett.* **39**, 1933 (2014).
51. H. Zhang, H. Xiao, P. Zhou, X. Wang, and X. Xu, *Opt. Express* **22**, 10248 (2014).
52. E. M. Dianov, I. A. Bufetov, V. M. Mashinsky, V. B. Neustruev, O. I. Medvedkov, A. V. Shubin, M. A. Melkumov, A. N. Gur'yanov, V. F. Khopin, and M. V. Yashkov, *Quantum Electron.* **34**, 695 (2004).
53. M. Duhant, W. Renard, G. Canat, T. N. Nguyen, F. Smektala, J. Troles, Q. Coulombier, P. Toupin, L. Brilland, P. Bourdon, and G. Renversez, *Opt. Lett.* **36**, 2859 (2011).
54. C. Antonelli, A. Mecozzi, and M. Shtaif, *Opt. Lett.* **38**, 1188 (2013).
55. J. Zhou, *Opt. Express* **22**, 21393 (2014).
56. E. N. Christensen, J. G. Koefoed, S. M. M. Friis, M. A. U. Castaneda, and K. Rottwitt, *Sci. Rep.* **6**, 34693 (2016).
57. M. Esmaeelpour, R. Ryf, N. K. Fontaine, H. Chen, A. H. Gnauck, R.-J. Essiambre, J. Toulouse, Y. Sun, and R. Lingle, Jr., *J. Lightwave Technol.* **34**, 1864 (2016).
58. D. J. Richardson, J. M. Fini, and L. E. Nelson, *Nat. Photonics* **7**, 354 (2013).
59. R. G. H. van Uden, R. Amezcua Correa, E. Antonio Lopez, F. M. Huijskens, C. Xia, G. Li, A. Schülzgen, H. de Waardt, A. M. J. Koonen, and C. M. Okonkwo, *Nat. Photonics* **8**, 865 (2014).
60. G. Li, N. Bai, N. Zhao, and C. Xia, *Adv. Opt. Photonics* **6**, 413 (2014).
61. M. N. Islam, *IEEE J. Sel. Top. Quantum Electron.* **8**, 548 (2002).
62. J. C. Bouteiller, K. Brar, J. Bromage, S. Radic, and C. Headley, *IEEE Photonics Technol. Lett.* **15**, 212 (2003).
63. J. Li, J. Du, L. Ma. M.-J. Li, K. Xu, and Z. He, *Opt. Express* **25**, 810 (2017).
64. M. F. Ferreira, J. F. Rocha, and J. L. Pinto, *Opt. Quantum Electron.* **26**, 35 (1994).
65. N. A. Olsson and J. P. van der Ziel, *Appl. Phys. Lett.* **48**, 1329 (1986).
66. N. A. Olsson and J. P. van der Ziel, *Electron. Lett.* **22**, 488 (1986).
67. A. Loayssa, D. Benito, and M. J. Garde, *Opt. Fiber Technol.* **8**, 24 (2002).
68. M. J. LaGasse, W. Charczenko, M. C. Hamilton, and S. Thaniyavarn, *Electron. Lett.* **30**, 2157 (1994).
69. K. J. Williams and R. D. Esman, *Electron. Lett.* **30**, 1965 (1994).
70. T. Tanemura, Y. Takushima, and K. Kikuchi, *Opt. Lett.* **27**, 1552 (2002).
71. Y. Shen, X. Zhang, and K. Chen, *Proc. SPIE* **5625**, 109 (2005).
72. A. Wiberg and P. O. Hedekvist, *Proc. SPIE* **5466**, 148 (2004).
73. T. Schneider, M. Junker, and D. Hannover, *Electron. Lett.* **40**, 1500 (2004).
74. L. Xing, L. Zhan, S. Luo, and Y. Xia, *IEEE J. Quantum Electron.* **44**, 1133 (2008).
75. K. Shiraki, M. Ohashi, and M. Tateda, *J. Lightwave Technol.* **14**, 50 (1996).
76. A. Yeniay, J.-M. Delavaux, and J. Toulouse, *J. Lightwave Technol.* **20**, 1425 (2002).
77. G. Qin, H. Sotobayashi, M. Tsuchiya, A. Mori, T. Suzuki, and Y. Ohishi, *J. Lightwave Technol.* **26**, 492 (2008).
78. K. S. Abedin, *Opt. Express* **13**, 10266 (2005).
79. R. W. Boyd, *Nonlinear Optics*, 2nd ed., Academic Press, San Diego, CA, Chap. 9 (2003).
80. G. P. Agrawal, *Nonlinear Fiber Optics*, 3rd ed., Academic Press, San Diego, CA, Chap. 9 (2001).
81. C. L. Tang, *J. Appl. Phys.* **37**, 2945 (1966).
82. N. A. Olsson, and J. P. van der Ziel, *J. Lightwave Technol.* **5**, 147 (1987).
83. J. H. Lee, T. Tanemura, K. Kikuchi, T. Nagashima, T. Hasegawa, S. Ohara, and N. Sugimoto, *Opt. Lett.* **30**, 1698 (2005).

84. K. S. Abedin, *Opt. Lett.* **31**, 1615 (2006).
85. K. S. Abedin, *Opt. Express* **14**, 11766 (2006).
86. R. W. Tkach and A. R. Chraplyvy, *Opt. Quantum Electron.* **21**, S105 (1989).
87. C. G. Atkins, D. Cotter, D. W. Smith, and R. Wyatt, *Electron. Lett.* **22**, 556 (1986).
88. D. Cotter, D. W. Smith, C. G. Atkins, and R. Wyatt, *Electron. Lett.* **22**, 671 (1986).
89. Y. Dong, T. Jiang, L. Teng, H. Zhang, L. Chen, X. Bao, and Z. Lu, *Opt. Lett.* **39**, 2967 (2014).
90. V. L. Iezzi, S. Loranger, M. Marois, and R. Kashyap, *Opt. Lett.* **39**, 857 (2014).
91. S. Preussler and T. Schneider, *Opt. Express* **23**, 26879 (2015).
92. Y. Stern, K. Zhong, T. Schneider, R. Zhang, Y. Ben-Ezra, M. Tur, and A. Zadok, *Photonics Res.* **2**, B18 (2014).
93. A. Choudhary, I. Aryanfar, S. Shahnia, B. Morrison, K. Vu, S. Madden, B. Luther-Davies, D. Marpaung, and B. J. Eggleton, *Opt. Lett.* **41**, 436 (2016).
94. A. R. Chraplyvy and R. W. Tkach, *Electron. Lett.* **22**, 1084 (1986).
95. D. J. Gauthier, A. L. Gaeta, and R. W. Boyd, *Photon. Spectra* **40**, 44 (2006).
96. R. W. Boyd, D. J. Gauthier, and A. L. Gaeta, *Optics Photon. News* **17**, 19 (2006).
97. F. Xia, L. Sekaric, and Y. Vlasov, *Nature Photon.* **1**, 65 (2007).
98. E. Mateo, F. Yaman, and G. Li, *Opt. Lett.* **33**, 488 (2008).
99. J. Liu, T. H. Cheng, Y. K. Yeo, Y. Wang, L. Xue, W. Rong, L. Zhou, G. Xiao, D. Wang, and X. Yu, *J. Lightwave Technol.* **27**, 1279 (2009).
100. Y. Okawach, M. Bigelow, J. Sharping, Z. Zhu, A. Schweinsberg, D. J. Gauthier, R. W. Boyd, and A. L. Gaeta, *Phys. Rev. Lett.* **94**, 153902 (2005).
101. C. Jauregui, H. Ono, P. Petropoulos, and D. J. Richardson, *Conf. Optical Fiber Commun.* (OFC 2006), paper PDP2 (2006).
102. K. Y. Song, K. S. Abedin, K. Hotate, M. G. Herráez, and L. Thévenaz, *Opt. Express* **14**, 5860 (2006).
103. M. E. Marhic, N. Kagi, T.-K. Chiang, and L. G. Kazovsky, *Opt. Lett.* **21**, 573 (1996).
104. M. Karlsson, *J. Opt. Soc. Am. B* **15**, 2269 (1998).
105. J. Hansryd and P. A. Andrekson, *IEEE Photon. Technol. Lett.* **13**, 194 (2001).
106. M. Westlund, J. Hansryd, P. A. Andrekson, and S. N. Knudsen, *Electron. Lett.* **38**, 85 (2002).
107. J. Hansryd, P. A. Andrekson, M. Westlund, J. Li, and P. O. Hedekvist, *IEEE J. Sel. Top. Quantum Electron.* **8**, 506 (2002).
108. V. Gordienko, M. F. C. Stephens, A. E. El-Taher, and N. J. Doran, *Opt. Express* **25**, 4810 (2017).
109. Z. Tong, C. Lundstrom, P. A. Andrekson, M. Karlsson, and A. Bogris, *IEEE J. Sel. Topics Quantum Electron.* **18**, 1016 (2012).
110. M. F. C. Stephens, M. Tan, I. D. Phillips, S. Sygletos, P. Harper, and N. J. Doran, *Opt. Express* **22**, 11840 (2014).
111. A. D. Ellis, M. Tan, M. A. Iqbal, M. A. Z. Al-Khateeb, V. Gordienko, G. S. Mondaca, S. Fabbri, M. F. C. Stephens, M. E. McCarthy, A. Perentos, I. D. Phillips, D. Lavery, G. Liga, R. Maher, P. Harper, N. Doran, S. K. Turitsyn, S. Sygletos, and P. Bayvel, *J. Lightwave Technol.* **34**, 1717 (2016).
112. M. E. Marhic, Y. Park, F. S. Yang, and L. G. Kazovsky, *Opt. Lett.* **21**, 1354, (1996).
113. K. K. Y. Wong, M. E. Marhic, K. Uesaka, and L. G. Kazovsky, *IEEE Photon. Technol. Lett.* **14**, 911 (2002).
114. C. J. McKinstrie and S. Radic, *Opt. Lett.* **27**, 1138 (2002).
115. C. J. McKinstrie, S. Radic, and A. R. Chraplyvy, *IEEE J. Sel. Top. Quantum Electron.* **8**, 538 (2002).
116. S. Radic and C. J. McKinstrie, *Opt. Fib. Technol.* **9**, 7 (2003).
117. S. Radic, C. J. Mckinstrie, A. R. Chraplyvy, G. Raybon, J. C. Centanni, C. G. Jorgensen, K. Brar, and C. Headley, *IEEE Photon. Technol. Lett.* **14**, 1406 (2002).

118. M. Yu, C. J. McKinstrie, G. P. Agrawal, *Phys. Rev.* **48**, 2178 (1993).
119. R. H. Stolen and J. E. Bjorkholm, *IEEE J. Quantum Electron.* **18**, 1062 (1982).
120. K. Inoue, *J. Lightwave Technol.* **12**, 1916 (1994).
121. R. M. Jopson and R. E. Tench, *Electron. Lett.* **29**, 2216 (1993).
122. O. Leclerc, B. Lavigne, E. Balmefrezol et al., *J. Lightwave Technol.* **21**, 2779 (2003).
123. M. F. C. Stephens, M. Tan, V. Gordienko, P. Harper, and N. J. Doran, *Opt. Express* **25**, 24312 (2017).
124. M. F. C. Stephens, V. Gordienko, and N. J. Doran, *Opt. Express* **25**, 10597 (2017).

7 All-Optical Switching

7.1 INTRODUCTION

An optical switch with fully transparent features in both time and wavelength domains is a key device for providing several functions required in optical signal processing. A practical optical switch should have a broad bandwidth over the entire transmission band and should be capable of ultrahigh-speed operation at a data rate of 100 Gb/s or higher. Some of the first demonstrations of all-optical switching made use of in-line semiconductor optical amplifiers (SOAs). Different nonlinear phenomena inside a SOA were considered, such as cross-polarization modulation, cross-gain modulation, and four-wave mixing (FWM) [1–3]. However, since the performance of a SOA depends on the carrier dynamics, its operation is limited to some tens of gigahertz. Ultrafast switching operations with speeds beyond the limits of electrical devices can only be achieved through an all-optical approach. The third-order nonlinearity in optical fibers can provide such features, especially if highly nonlinear fibers are considered. Optical switching can be achieved based on self-phase modulation (SPM), cross-phase modulation (XPM) and FWM effects in optical fibers [4–14]. In particular, all-optical switching approaches using parametric amplification both in phase insensitive (PI) and in phase sensitive (PS) regimes have been reported [10–14].

7.2 SPM-INDUCED OPTICAL SWITCHING

SPM-induced optical switching can be realized using a Mach-Zehnder interferometer (MZI), as represented in Figure 7.1. The first coupler splits the input signal into two parts, which are recombined in the second coupler. In general, the two arms of the interferometer can have different lengths and propagation constants. The two couplers can also have different power-splitting fractions, f_1 and f_2. Considering a continuous wave (CW) input beam with power $P_0 = |A_0|^2$ incident at one input port, the amplitudes of the optical fields at the two output ports of the first coupler are given by:

$$A_1 = \sqrt{f_1} A_0, \quad A_2 = i\sqrt{1-f_1} A_0 \tag{7.1}$$

During the propagation along the interferometer arms, both fields acquire both linear and SPM-induced phase shifts. As a result, the two fields reaching the second coupler take the following form:

$$A'_1 = A_1 \exp(i\beta_1 L_1 + if_1 \gamma P_0 L_1) \tag{7.2}$$

$$A'_2 = A_2 \exp(i\beta_2 L_2 + i(1-f_1)\gamma P_0 L_2), \tag{7.3}$$

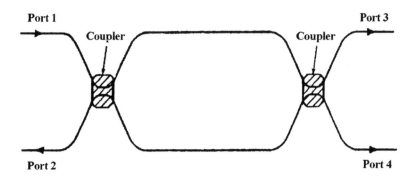

FIGURE 7.1 Schematic illustration of a Mach-Zehnder interferometer.

where L_1 and L_2 are the lengths, and β_1 and β_2 are the propagation constants for the two arms of the interferometer.

The optical fields at the output ports of the MZI can be obtained by using the transfer matrix of the second fiber coupler and are given by:

$$\begin{pmatrix} A_3 \\ A_4 \end{pmatrix} = \begin{pmatrix} \sqrt{f_2} & i\sqrt{1-f_2} \\ i\sqrt{1-f_2} & \sqrt{f_2} \end{pmatrix} \begin{pmatrix} A'_1 \\ A'_2 \end{pmatrix} \quad (7.4)$$

Using Eqs. (7.1)–(7.4), the transmittivity $T_3 = |A_3|^2 / |A_0|^2$ through the bar port of the MZI is given by:

$$T_3 = f_1 f_2 + (1-f_1)(1-f_2) - 2\left[f_1 f_2 (1-f_1)(1-f_2)\right]^{1/2} \cos(\Delta\phi_L + \Delta\phi_{NL}) \quad (7.5)$$

where

$$\Delta\phi_L = \beta_1 L_1 - \beta_2 L_2, \quad \Delta\phi_{NL} = \gamma P_0 \left[f_1 L_1 - (1-f_1)L_2\right] \quad (7.6)$$

are the linear and nonlinear parts of the relative phase shift.

The result in Eq. (7.5) becomes particularly simple if the MZI is made using two 3-dB couplers, so that $f_1 = f_2 = 1/2$. In this case, the transmittivity is given by:

$$T = \sin^2\left[(\Delta\phi_L + \Delta\phi_{NL})/2\right] \quad (7.7)$$

As long as the two arms of the interferometer have different lengths, the SPM-induced phase shift allows switching the output from one of the ports from low to high (or vice versa) by changing the input peak power of the incident signal.

Another scheme to realize SPM-based optical switching makes use of the Sagnac interferometer, as represented in Figure 7.2. It is made by connecting the two outputs of a fiber coupler through a piece of long fiber. The input beam is divided in the fiber coupler into two beams counterpropagating in the fiber loop. After one round trip,

All-Optical Switching

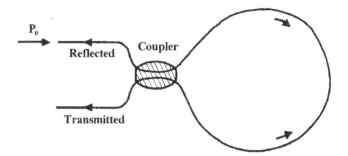

FIGURE 7.2 Schematic illustration of an all-fiber Sagnac interferometer.

both beams arrive at the same time at the coupler, where they interfere coherently, according to their relative phase difference. If a 3-dB fiber coupler is used, the input beam is totally reflected and the Sagnac interferometer serves as an ideal mirror. For this reason, it is also known as a nonlinear optical loop mirror (NOLM).

The amplitudes of the forward- and backward-propagating fields at the two output ports of the coupler are given by Eq. (7.1), with $A_f = A_1$ and $A_b = A_2$. After one round trip, both fields acquire linear and nonlinear phase shifts, taking the form:

$$A'_f = A_f \exp\left[\beta L + i\gamma\left(|A_f|^2 + 2|A_b|^2\right)L\right] \tag{7.8}$$

$$A'_b = A_b \exp\left[\beta L + i\gamma\left(|A_b|^2 + 2|A_f|^2\right)L\right] \tag{7.9}$$

where L is the loop length and β is the propagation constant.

The reflected and transmitted fields can be obtained by using the transfer matrix of the fiber coupler and are given by:

$$\begin{pmatrix} A_t \\ A_r \end{pmatrix} = \begin{pmatrix} \sqrt{f} & i\sqrt{1-f} \\ i\sqrt{1-f} & \sqrt{f} \end{pmatrix} \begin{pmatrix} A'_f \\ A'_b \end{pmatrix} \tag{7.10}$$

The transmittivity $T = |A_t|^2/|A_0|^2$ of the Sagnac loop is given by:

$$T = 1 - 4f(1-f)\cos^2\left[\left(f - \frac{1}{2}\right)\gamma P_0 L\right] \tag{7.11}$$

From Eq. (7.11) we see that $T = 0$ for a 3-dB coupler (the loop acts as a perfect mirror). However, if $f \neq 0.5$, the fiber loop exhibits different behavior at low and high powers and can act as an optical switch. At low input powers, we have $T \approx 1 - 4f(1-f)$, and little light is transmitted if $f \approx 0.5$. At high input powers, we can have $T = 1$ if the condition

$$\left|f-\frac{1}{2}\right|\gamma P_0 L = (2N-1)\frac{\pi}{2}, \quad (N \text{ integer}) \tag{7.12}$$

is satisfied. For $N = 1$, a fiber loop 50-m-long will need a switching power of 63 W when $f = 0.45$ and $\gamma = 10 \text{ W}^{-1}/\text{km}$.

The switching power of a Sagnac interferometer can be reduced to levels below 1 mW if a fiber amplifier is included within the loop. This device is referred to as the *nonlinear amplifying loop mirror* (NALM), and it can operate as a switch even if a 3-dB coupler ($f = 0.5$) is used. In the NALM, the asymmetry is provided by the fiber amplifier, which is placed closer to one of the coupler arms than the other. In this case, light traveling clockwise around the loop has a different intensity to light traveling anticlockwise, and hence experience different amounts of SPM in the loop due to fiber nonlinearity. If the clockwise wave is amplified by a factor G, the transmittivity becomes:

$$T = 1 - 4f(1-f)\cos^2\left[\left(\frac{Gf+f-1}{2}\right)\gamma P_0 L\right] \tag{7.13}$$

In this case, we can have $T = 1$ if the condition

$$\left|\frac{Gf+f-1}{2}\right|\gamma P_0 L = (2N-1)\frac{\pi}{2} \tag{7.14}$$

Assuming $f = 0.5$ and $N = 1$, Eq. (7.14) provides the following result for the switching power:

$$P_0 = \frac{2\pi}{(G-1)\gamma L} \tag{7.15}$$

It can be seen that the switching power is reduced by a factor up to 1000 for an amplification factor $G \approx 30$ dB. Using a 4.5 m-long fiber amplifier, which provided a gain of only 6 dB, a switching power of 0.9 W was reported for a 306-m fiber loop using a 3-dB coupler [15].

Using short pulses in Sagnac interferometers, the power dependence of loop transmittivity, as given by Eq. (7.11), can lead to significant distortion and pulse narrowing [16]. This happens since only the central part of the pulse is sufficiently intense to undergo switching. The same feature can be used for pulse shaping and pulse cleanup. For example, a low-intensity pedestal accompanying a short optical pulse can be removed by passing it through a Sagnac interferometer [17].

Nonlinear switching without the deformation effect can be achieved by using optical solitons, since they have a uniform nonlinear phase across the entire pulse. Any low-energy dispersive radiation accompanying a train of optical solitons is reflected back, while they are transmitted by the loop. As seen in Chapter 3, optical solitons can be observed by launching ultrashort pulses at a wavelength in the anomalous-dispersion regime of the fiber. Soliton switching in Sagnac interferometers has been observed experimentally by several authors [18,19].

7.3 XPM-INDUCED OPTICAL SWITCHING

As discussed in Section 2.1.3, the effect of XPM produces an alteration in the phase of one pulse due to the intensity of another pulse propagating at the same time in the fiber. This effect can be used in a Sagnac interferometer, as shown in Figure 7.2, to realize an optical switch. Let us consider a control pulse that is injected into the Sagnac loop such that it propagates in only one direction. Due to the XPM effect in the fiber, the signal pulse that propagates in that direction experiences a phase shift while the counterpropagating pulse remains unaffected. From Eq. (2.28), the XPM-induced phase shift is given by $\Delta\phi_{NL} = 2\gamma P L_{eff}$, where P is the power of the control pulse. If this power is such that $\Delta\phi_{NL} = \pi$, the signal pulse will be transmitted instead of being reflected. The potential of XPM-induced switching in all-fiber Sagnac interferometers was demonstrated in several experiments [5–8].

Sagnac interferometers can be used for demultiplexing optical time-division multiplexed (OTDM) signals. In this case, the control signal consists of a train of optical pulses that is injected into the loop such that it propagates in a given direction. If the control (clock) signal is timed such that it overlaps with pulses belonging to a specific OTDM channel, the XPM-induced phase shift allows the transmission of this channel, whereas all the remaining channels are reflected. Different channels can be demultiplexed simultaneously by using several Sagnac loops [20]. A 11-km-long Sagnac loop was used in 1993 to demultiplex a 40 Gb/s signal to individual 10-Gb/s channels [21].

The main limitation of a Sagnac interferometer used to provide XPM-induced switching and demultiplexing stems from the weak fiber nonlinearity. Fibers with a length of several kilometers are necessary for a control pulse power in the mW range. In these circumstances, if the signal and control pulses have different wavelengths, the walk-off effects due to the group-velocity mismatch must be taken into consideration. The control pulse width and the walk-off between the signal and control pulses determine the switching speed as well as the maximum bit rate for demultiplexing with a NOLM [22].

A walk-off-free NOLM was proposed that consisted of nine sections of short dispersion-flattened fibers with different-dispersion values [8]. However, such a device has a very complex configuration. A simpler approach to suppress the walk-off effect consists of using a fiber whose zero-dispersion wavelength (ZDW) lies between the signal and the control wavelengths, such that both experience the same group velocity. A Sagnac loop interferometer operating under these conditions was realized in 1990 using a 200-m-long polarization-maintaining fiber [23]. This Sagnac loop used 120-ps pump pulses with 1.8-W peak power at 1320 nm to switch the 1540-nm signal. In a 2002 experiment, a 100-m long highly nonlinear dispersion-shifted fiber (HNL-DSF) with a nonlinear parameter $\gamma = 15\ W^{-1} km^{-1}$ and a ZDW at 1553.5 nm was used for error-free demultiplexing of 320 Gb/s TDM-signals down to 10 Gb/s [24].

Another possibility to suppress the pulse walk-off is to use an orthogonally polarized pump at the same wavelength as that of the signal. In this case, the group-velocity mismatch due to the polarization-mode dispersion is relatively low. Moreover, it can be used to advantage by constructing a Sagnac loop consisting of multiple sections of polarization-maintaining fibers that are spliced together in such a way that the slow and fast axis are interchanged periodically. As a result, the pump and signal pulses are forced to collide multiple times inside the Sagnac loop, and the XPM-induced phase shift is enhanced significantly. This idea has been verified using two orthogonally polarized pump and signal soliton pulses, which were launched in 10.2-m loop constituted by 11 sections [25].

Both the interaction length and the control power can be considerably decreased using highly nonlinear microstructured fibers. In a 2002 experiment, XPM-induced switching of 2.6 ps (full-width at hal-maximum) signal pulses in a Sagnac interferometer built with a 5.8-m-long microstructured fiber was achieved [26]. Optical time-division demultiplexing of 160-Gb/s OTDM signal based on XPM and subsequent optical filtering has also been demonstrated more recently using a 2-m-long Bi_2O_3-based highly nonlinear microstructured fiber with $\gamma > 600$ W^{-1}/km and GVD ~50 ps^2/km [27]. Using a such fiber, the long-term stability was drastically improved as compared with silica-fiber devices because the very short fiber length reduces the phase drift between the signal and control pulses.

By launching a CW beam together with control pulses at a different wavelength in a balanced Sagnac interferometer, simultaneous XPM-induced optical switching and wavelength conversion are realized. The CW beam is reflected in the absence of control pulse, while a slice of it is transmitted when a control pulse is present. As a result, a pulse train at the CW wavelength is produced. In one experiment, a train of highly chirped 60-ps control pulses at 1533 nm was used to convert the 1554-nm CW radiation into a high-quality pulse train [28].

Gaussian or sech pulses occupy a relatively large bandwidth due to their slow spectral decay in the tail, which limits the spectral efficiency of the transmission system. In order to overcome this bottleneck, an optical Nyquist pulse and its OTDM transmission (Nyquist OTDM) has been proposed [29]. The Nyquist pulse spectrum can be confined within a finite bandwidth, which provides a substantial advantage in terms of higher spectral efficiency [30] and increased tolerance to chromatic dispersion and polarization-mode dispersion even at an ultrafast symbol rate [31], as well as the efficient add-drop multiplexing of OTDM-WDM signals [32]. In a 2015 experiment, the 640 to 40 Gbaud demultiplexing of differential phase-shift keying (DPSK) and differential quadrature phase-shift keying (DQPSK) Nyquist OTDM signals was successfully demonstrated in an all-optical NOLM switch with a greatly improved bit-error rate (BER) performance compared with conventional Gaussian control pulses [33].

XPM-induced optical switching can also be realized using an MZI. If both branches have the same properties and in the absence of any control pulse, the two partial pulses created in the first fiber coupler experience the same phase shift, and they will interfere constructively in the second fiber coupler of the interferometer. However, if a control pulse at a different wavelength is injected into one branch of the interferometer, it would change the signal phase through XPM in that arm and switch it.

If an OTDM signal is introduced at the interferometer input and a control (clock) signal is injected into one of its branches, a specific OTDM channel can be demultiplexed. Demultiplexers based on XPM-induced phase shift have attracted a special attention, since they present some advantages relative to SPM-based devices.

All-Optical Switching 97

In particular, multiple MZIs can be cascaded as remaining channels appear at the output end of MZI, rather than being reflected.

Considering a typical single mode fiber with a nonlinear parameter $\gamma = 1.3$ W^{-1}/km and an effective length $L_{eff} = 1$ km, a phase shift $\Delta\phi_{NL} = \pi$ requires an input power of 1.2 W for the control pulse. Reducing the effective core area A_{eff} increases the nonlinear parameter γ and, consequently, reduces the power required for XPM-induced switching. For example, using fibers with $A_{eff} = 2$ μm^2 in each arm of an MZI, an XPM-induced phase shift of 10° was achieved with a pump power of only 15 mW [34].

A scheme similar to that shown in Figure 8.2, concerning XPM-based wavelength conversion, can also be used for time-domain demultiplexing [35]. In this case, the probe with wavelength λ_1 corresponds to the OTDM data signal, whereas the pump at wavelength λ_2 corresponds to the clock signal. The clock pulses induce a spectral broadening of only those OTDM pulses that overlap with them in the time domain. An optical filter placed at the output selects these pulses, resulting in a demultiplexed channel at the clock wavelength.

Multiple channels can be simultaneously demultiplexed by using multiple control pulses at different wavelengths. In a 2002 experiment, this technique has been employed to demultiplex four 10-Gb/s channels from a 40-Gb/ OTDM signal using a 500-m-long highly nonlinear fiber [36].

Using shorter fiber lengths helps with the problem of group-velocity mismatch between the signal and clock pulses. This is possible if microstructured fibers or non-silica fibers made of materials with a higher nonlinearity than silica are employed. In a 2006 experiment, a 2-m-long bismuth-oxide fiber has been employed to demultiplex a 160 Gb/s bit stream using control pulses at a 10-GHz repetition rate [37]. Such demultiplexer exhibited little polarization sensitivity, since a high-speed polarization scrambler was employed to randomize the state of polarization (SOP) of the input bit stream.

XPM-induced optical switching can also be realized with a Kerr shutter, consisting of an optical fiber and a polarizer. Both the signal and the control pulses are linearly polarized, making an angle of 45° between their polarization directions. The polarizer at the fiber output is adjusted perpendicularly to the polarization of the input signal. As a consequence, no signal light can pass the polarizer in the absence of a control pulse.

The Kerr shutter makes use of the XPM-induced nonlinear birefringence. In the presence of the control pulse, a slower axis is induced parallel to its polarization direction, which determines a rotation of the signal sate of polarization. A fraction of the signal pulse can now pass through the polarizer, depending on the intensity of the control pulse.

Since the signal and control pulses have different wavelengths, the performance of a Kerr shutter used for optical switching depends on group-velocity mismatch between them. This mismatch can be reduced by using highly nonlinear fibers, since much shorter lengths are needed then. In a 2005 experiment, only a 1-m-long bismuth-oxide fiber was employed to realize a Kerr shutter [38]. This device enabled the demultiplexing of a 160-Gb/s bit stream, using a train of 3.5-ps control pulses at a 10-GHz repetition rate.

7.4 OPTICAL SWITCHING USING FWM

FWM can also be used to switch a given sequence of signal pulses propagating along an optical fiber. To realize such operation, a control signal consisting of a train of optical pulses with a frequency (ω_c) different from that of the signal pulses (ω_s) is also injected into the fiber. When a signal and control pulse are present together in the fiber, a switched signal appears at the idler frequency, ω_i, given by

$$\omega_i = 2\omega_c - \omega_i \tag{7.16}$$

Using a filter at the fiber output, the switched signal can be isolated from the control and original signal sequence.

Optical bandwidths of the order of 200 nm can be achieved with an FWM-based fiber switch, in spite of the limitations imposed by the phase-matching condition [39]. Such device can also be used as a demultiplexer [40]. In this case, the pump for the FWM process is provided by the clock signal. A pulse at the idler frequency appears only when a clock pulse overlaps with a signal pulse of the channel to be demultiplexed. In this way, a replica of the demultiplexed channel is obtained at the idler frequency. In a 1996 experiment, a 500-Gb/s bit stream was demultiplexed into 10-Gb/s channels using clock pulses of 1-ps-duration [41]. This scheme can also amplify the demultiplexed channel through parametric gain inside the same fiber [42].

The parametric gain depends on the relative SOP of the signal and pump waves. As a result, the polarization sensitivity becomes a problem affecting the performance of an FWM-based demultiplexer. A polarization-diversity technique was suggested in 2004 for solving this problem [43]. Figure 7.3 illustrates this scheme, which includes a short piece of polarization-maintaining fiber (PMF) before the highly nonlinear fiber used for the FWM. The control pulses are polarized at 45° with respect to the principal axes of the PMF, which also separates the randomly polarized signal pulses into two orthogonally polarized parts. This technique was used to demultiplex 10-Gb/s channels from a 160-Gb/s bit stream with a polarization sensitivity below 0.5 dB.

Another main problem associated with the conventional FWM-based switching is related with the different wavelengths of the original and of the switched signal sequences. A technique to solve this problem was proposed by Watanabe et al. in a

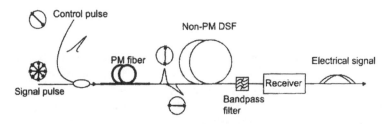

FIGURE 7.3 A polarization-insensitive FWM-based demultiplexer. (From T. Sakamoto et al., *IEEE Photon. Technol. Lett.* 16, 563, 2004. © 2004 IEEE. With permission [43].)

2008 experiment [9]. In such an approach, the optical signal wave E_S and the control (pump) wave E_P were launched into a nonlinear fiber. The fiber output was connected to a polarizer that transmits only linearly polarized light along its main axis. Using a polarization controller at the fiber input, the SOP of E_S was adjusted such that no light passes the polarizer in the absence of pump light. The SOP of E_P was also adjusted to linear polarization and aligned at ~45° with respect to the polarizer's main axis.

The configuration of the proposed fiber switch is similar to that of the conventional optical Kerr shutter, which is based on the XPM effect. However, in contrast to the Kerr shutter, the proposed switch has the advantage of using optical parametric amplification provided by FWM. To achieve the necessary phase matching, the pump wave was tuned to the ZDW, λ_{ZD}, of the fiber. The efficiency of FWM process is maximum when the two waves have the same SOP, whereas it is negligible when the SOPs of the two waves are orthogonal to each other. Therefore, decomposing E_S into two linearly polarized components, parallel ($E_{S\parallel}$) and orthogonal ($E_{S\perp}$) with respect to E_P, only $E_{S\parallel}$ is amplified by FWM. The output power at the output of the fiber optic parametric amplifier (FOPA) switch is

$$P_{S\text{-}out} = \frac{1}{2}\Delta P_{S\parallel} \tag{7.17}$$

where the factor 1/2 is due to the polarizer, and $\Delta P_{S\parallel}$ is the power increase of $E_{S\parallel}$ due to parametric amplification. The switching gain is approximately given by [9]:

$$G_S = \frac{P_{S\text{-}out}}{P_{S\text{-}in}} = \frac{1}{4}\exp(-\alpha L)\phi_{NL}^2 \tag{7.18}$$

where $P_{S\text{-}in}$ is the signal input power, α is the fiber loss coefficient, and $\phi_{NL} = \gamma P_P L_{eff}$ is the nonlinear phase shift.

Besides the fact that there is no frequency shift of the signal, the proposed FOPA switch provided also an extinction ratio of more than 30 dB because the unswitched component of E_S was eliminated by the polarizer. Such an extinction ratio could be improved by increasing the switching gain. Considering the ultrafast response of FWM in the fiber, the FOPA switch has a potential for application to signal processing at data rates of 1 Tb/s and higher.

A FOPA can be operated in a PI or in a PS mode. In the standard single-pump PI scheme, two interacting waves at different frequencies, one of high optical power (pump) and another of low optical power (signal), copropagate in a highly nonlinear fiber (HNLF). The FWM process produces a third wave (idler) at a frequency given by Eq. (7.16). As a result, there are three waves at the output of the PI with correlated phases. The relative phase among the three interacting waves is given by

$$\varphi_{rel} = 2\varphi_p - \varphi_s - \varphi_i, \tag{7.19}$$

with ϕ_p, ϕ_s and ϕ_i being the phases of the pump, signal and idler waves, respectively. In a PI amplifier, the idler wave is generated so that ϕ_{rel} is constant and the PI gain in the saturated regime is only dependent on the power of the interacting waves so that the output pump power depends only on the input signal power. This mechanism has been used by several authors [10–12]. In particular, all-optical switching using very small control pulse energies at a picosecond time scale was demonstrated in a 2008 experiment [10]. Switching with 3 dB of extinction using a control pulse energy of only 19 aJ (150 photons) was performed in this experiment using a standard HNLF. Using an appropriately engineered fiber, fewer photons were necessary to control the strong light signal in [11,12].

In a FOPA operating in a PS mode, at least three interacting and phase-correlated waves are required at the HNLF input [13,14,44]. In order to maximize the efficiency of the PS process, the signal and idler waves should have a nearly equal power. Operating the phase sensitive amplifier (PSA) in the small-signal (unsaturated) regime, the signal and idler waves are equally amplified and the pump power remains nearly constant. However, when the input powers of the signal and idler waves are sufficiently high, the PSA gain becomes saturated, leading to depletion of the pump wave. In this case, the output pump power is dependent on the power and phases of the interacting waves. By controlling the relative phase among the three interacting waves at the PSA input, it is possible to control the flow of power to or from the pump.

In a 2015 experiment [45], an all-optical switch using a PSA operated in the saturation regime was demonstrated. A 500-m long HNLF, with a ZDW at 1541.1 nm and a nonlinear coefficient $\gamma = 11.5$ W^{-1}/km was used in this experiment. By changing the phase of a 0.9 mW signal/idler pair wave by $\pi/2$ rad, a pump power extinction ratio of 30.4 dB (99.91%) has been achieved. This is shown in Figure 7.4a, where a maximum pump power of 28.4 dBm is observed in the small signal region, whereas a minimum pump power of −2 dBm is achieved when individual input signal and idler powers of −0.4 dBm are injected into the PSA. At the point of maximal depletion, many higher-order idlers were generated forming a comb-like spectrum, as illustrated in Figure 7.4b. Figure 7.4c shows that the signal and idler output powers increase linearly with the signal (and idler) input powers, both for the PI and PSA cases, until the saturation threshold is reached at around −4 dBm (−6 dBm). In the saturation region, the phase-matching condition is modified, and the energy flows from the pump to the higher-order idlers. Changing the relative phase of the signal and idler waves produced a variation of the pump, signal and idler powers at the PSA output as shown in Figure 7.4d. At the point of maximal pump power depletion, half of the pump power is transferred to the idler, around 20% goes to the signal, whereas the remaining pump power is distributed among the generated higher-order idlers. A periodic phase dependence of the pump power at the PSA output is observed in Figure 7.4d, and the maximum to minimum values are reached by changing the phase of the signal and idler waves $\pi/2$ rad (causing changes in the relative phase of π rad). This result illustrates very well the switching functionality performed by the PSA.

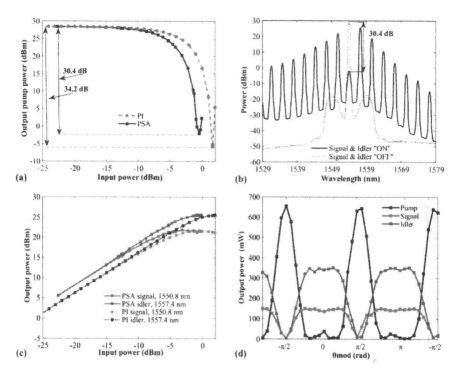

FIGURE 7.4 (a) Output pump power vs. input power, PI (dashed curve) and PSA (full curve); (b) optical spectra at the output of the PSA for an input signal and idler power of −0.4 dBm, and without any input wave; (c) input signal/idler power vs. output signal/idler power, PI and PSA cases; and (d) PSA output power vs. phase variation of the signal and idler waves. (From J. P. Cetina, et al., *Opt. Express*, 23, 33426, 2015. With permission of Optical Society of America [45].)

REFERENCES

1. K. E. Stubkjaer, *IEEE J. Sel. Top. Quantum Electron.* **6**, 1428 (2000).
2. S. A. Hamilton, B. S. Robinson, T. E. Murphy, S. J. Savage, and E. P. Ippen, *J. Lightwave Technol.* **20**, 2086 (2002).
3. H. Soto, J. C. Dominguez, D. Erasme, and G. Guekos, *Microw. Opt. Technol. Lett.* **29**, 205 (2001).
4. P. V. Mamyshev, *Eur. Conf. Opt. Commun.* (ECOC98), Madrid, Spain, pp. 475–477 (1998).
5. K. J. Blow, N. J. Doran, B. K: Nayar, and B. P. Nelson, *Opt. Lett.* **15**, 248 (1990).
6. M. Jino and T. Matsumoto, *Electron. Lett.* **27**, 75 (1991).
7. A. D. Ellis and D. A. Cleland, *Electron. Lett.* **28**, 405 (1992).
8. H. Bülow and G. Veith, *Electron. Lett.* **29**, 588 (1993).
9. S. Watanabe, F. Futami, R. Okabe, R. Ludwig, C. S. Langhorst, B. Huettl, C. Schubert, and H.-G. Weber, *IEEE J. Sel. Topics Quantum Electron.* **14**, 674 (2008).
10. P. A. Andrekson, H. Sunnerud, S. Oda, T. Nishitani, and J. Yang, *Opt. Express* **16**, 10956 (2008).

11. R. Nissim, A. Pejkic, E. Myslivets, B. P. Kuo, N. Alic, and S. Radic, *Science* **345**, 417 (2014).
12. A. Pejkic, R. R. Nissim, E. Myslivets, A. O. J. Wiberg, N. Alic, and S. Radic, *Opt. Express* **22**, 23512 (2014).
13. C. McKinstrie and S. Radic, *Opt. Express* **12**, 4973 (2004).
14. C. Lundström, B. Corcoran, M. Karlsson, and P. A. Andrekson, *Opt. Express* **20**, 21400 (2012).
15. M. E. Fermann, F. Harberl, M. Hofer, and H. Hochstrasser, *Opt. Lett.* **15**, 752 (1990).
16. N. J. Doran, D. S. Forrester, and B. K. Nayar, *Electron. Lett.* **25**, 267 (1989).
17. K. Smith, N. J. Doran, and P. G. J. Wigley, *Opt. Lett.* **15**, 1294 (1990).
18. K. J. Blow, N. J. Doran, and B. K. Nayar, *Opt. Lett.* **14**, 754 (1989).
19. M. N. Islam, E. R. Sunderman, R. H. Stolen, W. Pleibel, and J. R. Simpson, *Opt. Lett.* **14**, 811 (1989).
20. E. Bodtkrer and J. E. Bowers, *J. Lightwave Technol.* **13**, 1809 (1995).
21. D. M. Patrick, A. D. Ellis, and D. M. Spirit, *Electron. Lett.* **29**, 702 (1993).
22. K. Uchiyama, H. Takara, T. Morioka, S. Kawanishi, M. Saruwarari, *Electron. Lett.* **29**, 1313 (1993).
23. M. Jinno and T. Matsumoto, *IEEE Photon. Technol. Lett.* **2**, 349 (1990).
24. H. Sotobayashi, C. Sawagushi, Y. Koyamada, and W. Chujo, *Opt. Lett.* **27**, 1555 (2002).
25. J. D. Moores, K. Bergman, H. A. Haus, and E. P. Ippen, *Opt. Lett.* **16**, 138 (1991).
26. J. E. Sharping, M. Fiorentino, P. Kumar, and R. S. Windeler, *IEEE Photon. Technol. Lett.* **14**, 77 (2002).
27. K. Igarashi and K. Kikuchi, *IEEE J. Sel. Topics Quantum Electron.* **14**, 551 (2008).
28. R. A. Betts, J. W. Lear, S. J. Frisken, and P. S. Atherton, *Electron. Lett.* **28**, 1035 (1992).
29. M. Nakazawa, T. Hirooka, P. Ruan, and P. Guan, *Opt. Express* **20**, 112 (2012).
30. D. O. Otuya, K. Harako, K. Kasai, T. Hirooka, and M. Nakazawa, *Opt. Fiber Commun. Conf.* (OFC), paper M3G.2 (2015).
31. K. Harako, D. Seya, T. Hirooka, and M. Nakazawa, *Opt. Express* **21**, 21062 (2013).
32. H. N. Tan, T. Inoue, T. Kurosu, and S. Namiki, *Opt. Express* **21**, 20313 (2013).
33. T. Hirooka, D. Seya, K. Harako, D. Suzuki, and M. Nakazawa, *Opt. Express* **23**, 20858 (2015).
34. I. W. White, R. V. Penty, and R. E. Epworth, *Electron. Lett.* **24**, 340 (1988).
35. B. E. Olsson, and D. J. Blumenthal, *IEEE Photon. Technol. Lett.* **13**, 875 (2001).
36. L. Rau, W. Wang, B. E. Olsson, Y. Chiu, H. F. Chou, D. J. Blumenthal, and J. E. Bowers, *IEEE Photon. Technol. Lett.* **14**, 1725 (2002).
37. R. Salem, A. S. Lenihan, G. M. Carter, and T. E. Murphy, *IEEE Photon. Technol. Lett.* **18**, 2254 (2006).
38. J. H. Lee, T. Tanemura, T. Nagashima, T. Hasegawa, S. Ohara, N. Sugimoto, and K. Kikuchi, *Opt. Lett.* **30**, 1267 (2005).
39. M. C. Ho, K. Uesaka, M. Marhic, Y. Akasaka, and L. G. Kazovsky, *J. Opt. Lightwave. Technol.* **19**, 977 (2001).
40. P. A. Andrekson, N. A. Olsson, J. R. Olsson, J. R. Simpson, T. Tanbun-Ek, R. A. Logan, and M. Haner, *Electron. Lett.* **27**, 922 (1991).
41. T. Morioka, H. Takara, S. Kawanishi, K. Kitoh, and M. Saruwatari, *Electron. Lett.* **32**, 833 (1996).
42. J. Hansryd and P. A. Andrekson, *IEEE Photon. Technol. Lett.* 13, 732 (2001).
43. T. Sakamoto, K. Seo, K. Taira, N. S. Moon, and K. Kikuchi, *IEEE Photon. Technol. Lett.* **16**, 563 (2004).
44. R. Tang, J. Lasri, P. S. Devgan, V. Grigoryan, P. Kumar, and M. Vasilyev, *Opt. Express* **13**, 10483 (2005).
45. J. P. Cetina, A. Kumpera, M. Karlsson, and P. A. Andrekson, *Opt. Express* **23**, 33426 (2015).

8 Wavelength Conversion

8.1 INTRODUCTION

The possibility of generating signals at a desired frequency by all-optical wavelength conversion is extremely attractive for many different applications, including telecommunications, sensing, medicine, and defense. In particular, wavelength-conversion technologies are essential in future optical communication networks for improving the effective use of fiber transmission bandwidths and to solve the problem of network congestion caused by wavelength contention [1–3]. The capability of wavelength converting cheap and reliable optical sources into very different frequency bands as compared to the original is also very important [4,5].

Wavelength conversion has been performed using optical/electrical/optical converters with which the input optical signal is converted to an electrical signal by using photodetectors, electrical signal processing is performed, and finally, the electrical signal is used to modulate an optical source delivering a different wavelength. However, besides increasing the cost and the complexity of the network, these devices achieve limited transparency and typically introduce latency in a system. To improve the performance, all-optical wavelength converters (WCs) have been extensively investigated using nonlinear optical devices [6–12].

All-optical wavelength conversion can be realized using both semiconductor optical amplifier (SOA)-based converters [13,14] and fiber-based converters. WCs based on SOAs exhibit many attractive features but are generally limited to signal rates of ~40 Gb/s owing to carrier recombination lifetimes or intraband dynamics [15]. By contrast, converters based on nonlinear fiber effects can support almost unlimited bit rates and are transparent to the signal modulation format due to the femtosecond response time of the fiber nonlinearity [16]. Both cross-phase modulation (XPM) and four-wave-mixing (FWM) effects in fiber have been widely used to realize ultrahigh-speed wavelength conversion of return-to-zero data.

8.2 FWM-BASED WAVELENGTH CONVERTERS

Fiber-based FWM has been widely explored to perform wavelength conversion [17–20]. Besides being transparent to both data rate and modulation format, FWM can work in any spectral region as long as phase matching is satisfied. In a single-mode fiber, phase matching can be achieved either over a relatively broad spectral region around the pump wavelengths [20–22], or at narrow bands that can be far away from the pump wavelengths [22,23].

As discussed in Section 2.5, the FWM process generates an idler wave that is a replica of the signal wave, but at a different frequency. This effect can be used to realize the wavelength conversion. Considering the case of degenerate FWM, if the old frequency corresponds to the signal, the new frequency provided by the idler wave is given by:

$$f_i = 2f_p - f_s \tag{8.1}$$

In order to obtain the desired wavelength, we need to launch inside the fiber both the signal and a continuous wave (CW) pump with the appropriate wavelength. An optical filter placed at the output of the fiber blocks the pump and signal waves, while allowing the passage of the idler wave.

Assuming that the intensity of the pump wave is much higher than that of the signal and idler waves, Eqs. (2.36)–(2.39) with $U_1 = U_2 = U_p$, $U_3 = U_s$, and $U_4 = U_i$, given

$$\frac{\partial U_p}{\partial z} = i\gamma |U_p|^2 U_p \tag{8.2}$$

$$\frac{\partial U_s}{\partial z} = i2\gamma \left[|U_p|^2 U_s + U_p^2 U_i^* e^{-i\Delta k z} \right] \tag{8.3}$$

$$\frac{\partial U_i}{\partial z} = i2\gamma \left[|U_p|^2 U_i + U_p^2 U_s^* e^{-i\Delta k z} \right], \tag{8.4}$$

where γ is the nonlinear parameter, and $\Delta k = \beta_s + \beta_i - 2\beta_p$ is the phase mismatch of the wave vectors. As seen in Chapter 2, the effective phase mismatch with induced nonlinearity is given by

$$\kappa = \Delta k + 2\gamma P_p, \tag{8.5}$$

where P_p is the pump power. Using the results obtained in Section 6.4, the efficiency of the FWM converter can be written as

$$G_i = \frac{P_i(L)}{P_s(0)} = \left(\frac{\gamma P_p}{g} \right)^2 \sinh^2(gL), \tag{8.6}$$

where L is the fiber interaction length, and the parametric gain coefficient g is given by

$$g^2 = \left[(\gamma P_p)^2 - (\kappa/2)^2 \right] = -\Delta k \left(\frac{\Delta k}{4} + \gamma P_p \right). \tag{8.7}$$

The parameter g represents real gain over a conversion bandwidth corresponding to $-4\gamma P_p < \Delta k < 0$. The maximum gain for the parametric process is $g_{\max} = \gamma P_p$ and occurs if $\kappa = 0$ is satisfied. The conversion efficiency G_i can be much greater than 1 in such case. Thus, the FWM converter provides the amplification of a bit stream, while switching its wavelength, which is a very useful feature.

Wavelength Conversion

As seen in Section 2.5, the linear phase mismatch can be written as $\Delta k = \beta_2 \Omega_s^2$, where $\Omega_s = \omega_p - \omega_s$. Equation (8.5) shows that, for a given value of Ω_s, the phase mismatch is zero, and the conversion efficiency becomes maximum when the pump wavelength is in the anomalous dispersion regime, such that $\beta_2 = -2\gamma P_p / \Omega_s^2$. However, the conversion efficiency decreases if the signal wavelength deviates from the specific value of Ω_s. This determines the bandwidth over which wavelength conversion can be realized, which decreases for longer fiber lengths. A bandwidth larger than 80 nm can be achieved for fibers shorter than 100 m [24].

Fibers with a high nonlinear parameter γ will be advantageous for efficient wavelength conversion. On the other hand, in order to fulfill the phase-matching condition, fibers must be operated in their dispersion minimum. Fibers with a high value of γ and exhibiting the zero-dispersion wavelength (ZDW) around 1.5 μm are called highly nonlinear dispersion-shifted fibers (HNL-DSFs). In a 1998 experiment, a peak conversion efficiency of 28 dB was achieved over a bandwidth of 40 nm, using a 720-long DSF and a pulsed pump with a peak power of 600 mW [25]. Wavelength band conversion was also demonstrated in a 2001 experiment [26], through which the existing wavelength division multiplexing (WDM) sources in C-band were wavelength converted to the S-band. More than a 30-nm conversion bandwidth with greater than 4.7-dB conversion efficiency was measured in a 315-m long HNL-DSF by using 860-mW pump power at ~1532 nm [26].

The derivation of Eq. (8.6) ignores the pump depletion, fiber loss, competing nonlinear processes, and walk-off between the pump and the signal pulses. Since these effects would lower the parametric gain by lowering the pump power or reduce the interaction length, the conversion efficiency given by Eq. (8.6) corresponds effectively to an optimum value.

Compared to traditional fiber with one ZDW, microstructured optical fibers (MOFs) with two, three, and four ZDWs can be obtained through special design of the structural parameters of the fibers. Through FWM, two phase-matching sidebands can be generated in a fiber with one ZDW, four phase-matching sidebands in a PCF with two ZDWs, and six phase-matching sidebands in PCFs with three and four ZDWs [27–32]. Moreover, whereas a PCF with one or two ZDWs has only one anomalous dispersion region, there can be more anomalous dispersion regions in the case of PCFs with three and four ZDWs. This provides the possibility of achieving the wavelength conversion of optical solitons from an anomalous dispersion region to another [32].

When the group velocity dispersion (GVD) $|\beta_2|$ is small, we must take into account the fourth-order term in the expansion of Δk, which becomes

$$\Delta k = \beta_2 \Omega_s^2 + \frac{\beta_4}{12} \Omega_s^4 \qquad (8.8)$$

Equation (8.8) can be modified and written in terms of the involved wavelengths as

$$\Delta k = \left[2\pi c \left(\frac{1}{\lambda_p} - \frac{1}{\lambda_s} \right) \right]^2 \left[\beta_2 + \frac{1}{3} \beta_4 \pi^2 c^2 \left(\frac{1}{\lambda_p} - \frac{1}{\lambda_s} \right)^2 \right] \qquad (8.9)$$

For low pump powers, we can neglect the nonlinear term in Eq. (8.5), and the phase matching condition is given simply by $\Delta k = 0$. In order to realize this condition, Eq. (8.9) shows that the relation

$$\frac{1}{\lambda_p} - \frac{1}{\lambda_s} = \pm \frac{1}{\pi c} \sqrt{\frac{-3\beta_2}{\beta_4}} \qquad (8.10)$$

should be satisfied. As seen from Eq. (8.9), when $\beta_2 = 0$ the conversion bandwidth is proportional to $|\beta_4|^{-4}$. Hence, depending on the value of β_4, FWM-based wavelength conversion can be classified into three types [22]. The first one corresponds to conventional conversion and uses a highly nonlinear fiber (HNLF) with a moderate value $|\beta_4| \approx 1 \times 10^{-55}$ s^4/m. The second type corresponds to large values β_4 and provides narrowband conversion. The third one is realized using HNLFs with a small absolute value of β_4, which provides a broadband conversion. In a 2009 experiment, a FWM-based WC was implemented using a 100m-long HNLF with a nonlinear parameter $\gamma = 25$ W^{-1}/km and a reduced value $\beta_4 = 0.2 \times 10^{-55}$ s^4/m [22]. The pump wavelength was set to 1562.5 nm, slightly longer than the ZDW of 1562.1 nm, in order to have a small negative value of -1×10^{-29} s^2/m for β_2. The wavelength conversion in the reduced-β_4 HNLF can be realized within an extremely broad 3 dB-bandwidth of 222 nm from 1460 to 1682 nm. This result and other measurements obtained with different fiber lengths have confirmed that the bandwidths obtained for the reduced-β_4 HNLF are almost a factor 2 larger compared to those reported for conventional HNLFs [24,33].

In addition to the broadband wavelength conversion, there is also a growing interest in developing wavelength selective devices from multiple signals, such as reconfigurable optical add-drop multiplexers. In order to realize a selective wavelength conversion, Δk should be small in a narrow bandwidth around λ_s, as determined by Eq. (8.10). Therefore, besides β_2/β_4 being negative, we find from Eq. (7.9) that both $|\beta_2|$ and $|\beta_4|$ should be large. The desired value of β_2/β_4 and a tunable wavelength conversion can be realized by tuning the pump wavelength near the ZDW. This was confirmed in a 2009 experiment, using a 100-m-long HNLF with an enlarged value $\beta_4 = -2 \times 10^{-55}$ s^4/m and detuning the pump wavelength to a shorter wavelength from the ZDW of 1528.0 nm [22]. Tuning λ_p to 1527.2, 1527.0, and 1526.8, narrowband conversion was realized around 1600, 1610, and 1620 nm, respectively.

Broadband FWM-based WCs using phase-matching provided by positive fourth-order dispersion have enabled distant translation of continuous-wave [34] and/or amplitude modulated signals [35] from the near-infrared (NIR) to the short-wave infrared (SWIR) regions. However, for phase information translation into the SWIR frequency band, such broadband WCs designed using conventional HNLFs impose considerable limitations due to a broad out-of-(pump)-band amplified quantum noise generation. This limitation can be overcome by taking advantage of a narrowband WC relying on negative fourth-order dispersion. In a 2013 experiment, a dispersion fluctuation resilient HNLF with a negative value of β_4 was successfully used to demonstrate an error-free translation of 10 Gb/s phase-modulated information from the NIR to the SWIR (1.7–2.2 μm) band [36].

Actually, the SWIR and the mid-infrared (MIR) wavelength ranges deserve great interest due to the large number of potential applications that can be developed in these spectral regions, such as free-space communications, biological and chemical sensing, and spectroscopy. Moreover, 2 µm laser sources can be used for pumping nonlinear media to achieve all-optical ultrafast processing and generate light at targeted wavelengths deeper into the MIR through FWM.

Since silica glass presents too much loss beyond 2 µm, novel highly nonlinear materials, with high transparency over a broad wavelength range, have to be considered for operation of chalcogenide glass in the SWIR and MIR wavelength ranges. An approach relies on using microstructured fibers made of chalcogenide glass. This glass has a third-order nonlinearity up to 1000 times that of silica, as well as a broad transparency window up to 10 µm or 15 µm in the MIR, depending on the glass composition. In addition, as seen in Chapter 3, microstructured optical fibers offer great design flexibility. By controlling the structural parameters of a microstructured optical fiber (MOF), we can adjust its dispersion characteristics, which becomes especially important in the case of the FWM process. In a 2016 experiment, wavelength conversion in the 2 µm region by FWM in an AsSe and a GeAsSe chalcogenide MOF has been demonstrated [37]. A conversion efficiency of −25.4 dB was measured for 112 mW of coupled continuous wave pump in a 27-cm-long fiber.

A problem arising from the low dispersion in single-mode nonlinear systems is that it gives rise to unintended FWM-induced interactions between the input waves [38]. On the other hand, in order to realize the wavelength conversion of a specific signal wavelength, an accurate control of the fourth-order dispersion [39] is needed, which is difficult task in practice.

These and other limitations of the single-mode systems can be overcome if more than one spatial mode of the waveguide can be allowed to interact nonlinearly. This provides the opportunity for exploiting additional degrees of freedom: not only can frequencies in the same mode interact by mediation of $\chi^{(3)}$, but now each frequency and mode pair can be coupled nonlinearly to another such pair as part of an intermodal (IM) FWM interaction. An additional control over certain signal processing functionalities can be achieved in this way. In a 2017 experiment, IM FWM Bragg Scattering [40] has been used for wavelength conversion of optical signals from the C-band to the L-band using a few-mode fiber (FMF) [41]. More recently, a simultaneous threefold wavelength and modal conversion process of a 10-Gbit/s On/Off keying signal in a 1.8-km-long graded-index 6-LP-mode fiber has been demonstrated [42]. The principle of operation was based on a phase-matched IM FWM process occurring between the fundamental mode LP01 and three higher-order modes, LP11, LP02, and LP31 of a FMF.

In another 2019 experiment, IM FWM Bragg Scattering in a dispersive FMF was exploited to achieve selective wavelength conversion of individual WDM channels on a 100 GHz spacing [43]. The selectivity of parametric gain was obtained by dispersion design of the fiber such that the inverse group velocity (IGV) curves of the participating modes were parallel, and their dispersion was suitably large.

Figure 8.1 illustrates the principle of IM FWM Bragg Scattering used in the experiment of Ref. [43]. The pumps are in mode LP11 and have frequencies ω_{p1} and ω_{p2}, while the first of the two signals is in mode LP01 and has a frequency ω_s.

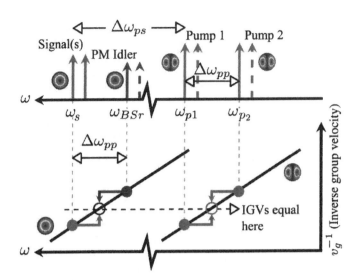

FIGURE 8.1 The wave configuration in which the pumps are in a higher order mode and the signal is in the fundamental mode. The modes and wavelengths are shown at the top, whereas the inverse group velocity curves and their relationship with phase matching are shown at the bottom. (From O. Anjum, et al., *Opt. Express*, 27, 24072, 2019. With permission of Optical Society [43].)

Then, a Bragg Scattering idler (PM Idler) arises at a frequency $\omega_{BSr} = \omega_s + \omega_{p2} - \omega_{p1}$. Phase matching is achieved for the PM Idler if the IGV v_g^{-1} evaluated at the average of the pump frequencies equals v_g^{-1} evaluated at the average of the two frequencies belonging to mode LP01 [44,45]. This is depicted by the horizontal dashed line in the bottom of Figure 8.1. Therefore, maximum idler generation at ω_{BSr} occurs when the signal is placed at a separation $\Delta\omega_{ps}$ from the pumps. Any other signals present in the vicinity are expected not to undergo efficient conversion. Moreover, phase matching can only hold when the pumps located at ω_{p1} and ω_{p2}, are in the LP11 mode.

If the IGV curves are *parallel*, then the signal to be converted can be selected by appropriately shifting the pump frequencies. For example, in order to convert the second signal (at the right of the first one) in Figure 8.1, the pump frequencies need to be shifted to those given by the dashed green lines. This will result in the idler generation at the dashed blue line being phase matched, whereas the first idler should now be far from phase matching.

Stimulated Brillouin scattering (SBS) has a threshold of only some milliwatts for long fibers, and this constitutes one of the main effects limiting the efficiency of FWM-based WCs. Since fiber optical parametric amplifiers (FOPAs) need pump powers around 1 W, some suitable technique to increase the SBS threshold becomes necessary. One commonly used technique for this purpose consists in modulating the phase of the pump wave, in order to increase its spectrum. However, this technique can introduce optical signal-to noise ratio (SNR) penalties in the case phase-modulated signals [46]. Moreover, it leads also to the broadening of the idler spectrum, which is clearly undesirable for WCs.

The use of two pumps, phase-modulated 180° out of phase can cancel out the broadening of the converted signal [47], while providing a nearly uniform gain over a wide bandwidth and polarization-independent operation of the converter. This can be understood by noting that the complex amplitude of the idler wave has the form $A_i \propto A_{p1}A_{p2}A_s^*$, where A_{p1} and A_{p2} are the pump amplitudes [16]. By modulating the pump waves in such a way that their phases remain equal but opposite in sign, the product $A_{p1}A_{p2}$ will not experience any modulation, and the idler spectrum will not be affected. Nearly polarization-independent wavelength conversion was achieved in a 2003 experience using a 1-km-long HNLF ($\gamma = 18$ W^{-1}/km) and binary phase-shift-keying modulation of the two orthogonally polarized pump waves [48]. The signal had a wavelength of 1557 nm, whereas the idler generated through FWM was near 1570 nm and exhibited the same bit pattern as the signal. A conversion efficiency of nearly 100% over a bandwidth of about 40 nm was achieved.

In a 2016 experiment, backward Raman amplification has been successfully applied to enhance the conversion efficiency in a two-orthogonal-pump FWM scheme [49]. The polarization insensitive feature was well preserved in the presence of the Raman pump. Wavelength conversion with ~0 dB efficiency and negligible polarization dependency was demonstrated by using a common 1-km long HNLF without pump dithering. The conversion efficiency was increased by ~29 dB with Raman enhancement.

An effective method to increase the SBS threshold consists in using short lengths of HNLFs to realize wavelength conversion. In a 2004 experiment, a 64-m-long silica MOF has been employed to implement an FWM-based WC [50], offering a reduction in fiber length by one order of magnitude. By 2005, only 1-m-length of a bismuth-oxide optical fiber higly nonlinear fiber (Bi-HNLF) with an ultra-high nonlinearity of ~1100 W^{-1} km^{-1} was used to make a FWM-based WC capable of operating at 80 Gb/s [51]. One main advantage of the Bi-HNLF is the possibility of fusion splicing to conventional silica fibers. More recently, a 35-cm-long Bi-HNLF was also used to demonstrate FWM-based wavelength conversion of 40 Gb/s polarization-multiplexed amplitude-shift keying (ASK) - differential phase shift keysing (DPSK) signals [52]. Using a dual-pump configuration, a conversion range greater than 30 nm was achieved. This experiment demonstrated the truly transparency of FWM wavelength conversion to modulation format. Signals in multilevel modulation format are particularly attractive in this context, since they allow the data to be transmitted at a higher bit rate than binary modulation without the need to increase the existing bandwidth of the electronic and optoelectronic components.

8.3 XPM-BASED WAVELENGTH CONVERTERS

One main limitation of an FWM-based converter comes from the fact that it is not possible to convert an unknown wavelength to a predetermined wavelength. This limitation is avoided using an XPM-based converter. A nonlinear optical loop mirror (NOLM) acting as a Sagnac interferometer can be used for this purpose. The data channel whose wavelength λ_2 needs to be converted is launched together with a CW whose wavelength λ_1 is chosen to coincide with the desired new wavelength. It affects the CW seed in only one direction, imposing an XPM-induced phase shift only in the time slots associated with the 1 bit. Such phase shift is converted into amplitude modulation by the interferometer, resulting in a bit pattern at the new wavelength λ_4, which is a replica of the original bit stream. In a 2000 experiment, a NOLM made using 3 km of a DSF was able to shift a 40 Gb/s data channel at 1547-nm by as much as 20 nm [53]. However, the required fiber length can be significantly reduced by employing highly nonlinear fibers.

Using an optical interferometer is not essential to realize an XPM-based WC. A simpler approach consists of using a DSF, in which a CW seed is launched along with the control signal. The XPM will act to broaden the spectrum of the CW lightwave, where a mark is copropagated with it. In this way, red- and blue-shifted sidebands are generated on the CW lightwave. By filtering out one of these sidebands, the wavelength-converted signals can be obtained [54–56]. The pulse-width of the converted pulse based on XPM in the DSF is determined by the power of the control pulse, dispersion effect, and walk-off time between the control pulse and the CW lightwave [53]. Broad-band pulse-width-maintained wavelength conversion can be realized by use of a short HNL-DSF because dispersion and walk-off effects can be reduced [56].

In a 2001 experiment [56], wavelength conversion of an 80-Gb/s data channel at the 1560-nm wavelength was realized using a 1-km-long fiber with $\gamma = 11\ W^{-1}/km$. The wavelength of the CW probe was varied in the range 1525–1554 nm, and a tunable optical filter with a 1.5-nm bandwidth was used at the fiber output to suppress the carrier and to select one of the sidebands, producing a wavelength-converted signal, which is a replica of the original bit stream. The pulse width of the converted signal remained nearly constant over a wide bandwidth.

The use of Raman gain can significantly enhance the performance of an XPM-based fiber WC [57,58]. Figure 8.2 shows a schematic of such Raman-gain-enhanced WC. It is basically constituted by a nonlinear fiber, a Raman pump laser, and a bandpass filter (BPF). The data pump light at wavelength λ_2 is combined with the CW probe light at wavelength λ_1 and injected into the fiber. The Raman pump amplifies simultaneously the pump light and the probe light and significantly enhances the

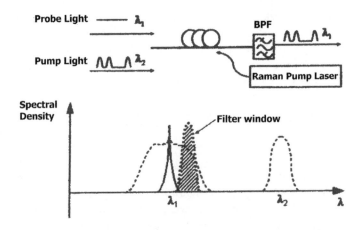

FIGURE 8.2 Schematic of an XPM-based Raman-gain-enhanced wavelength converter. (From W. Wang, et al., *J. Lightwave Technol.*, 23, 1105, 2005. © 2005 IEEE. With permission [57].)

amount of XPM. At the output of the fiber, a BPF is used to extract one of the sidebands on the CW probe, generating an amplitude-modulated signal from the phase modulation of the CW probe. This amplitude-modulated signal will thus form a wavelength-converted replica of the original data signal.

In the absence of Raman gain and considering the pulsed pump light and the CW probe light propagating in the fiber with the same polarization, the nonlinear equations that govern their behaviors can be written from Eqs. (2.23) and (2.24) as follows:

$$\frac{\partial U_s}{\partial z} + \frac{1}{v_{gs}} \frac{\partial U_s}{\partial t} = -\frac{\alpha}{2} U_s + i\gamma \left(|U_s|^2 + 2|U_p|^2 \right) U_s \qquad (8.11)$$

$$\frac{\partial U_p}{\partial z} + \frac{1}{v_{gp}} \frac{\partial U_p}{\partial t} = -\frac{\alpha}{2} U_p + i\gamma \left(|U_p|^2 + 2|U_s|^2 \right) U_p \qquad (8.12)$$

where U_j, $j = S, P$, is the slowly varying complex amplitude, v_{gj} is the group velocity, α is the attenuation coefficient, and γ is the fiber nonlinear parameter. In writing Eqs. (8.11) and (8.12), we have neglected both the GVD and the FWM effects. The solution of Eq. (8.11) at the output of the fiber with length L is given by

$$U_s(L,t) = U_s(0, t - L/v_{gs}) \exp(-\alpha L/2) \exp\{i\Delta\varphi(L,t)\} \qquad (8.13)$$

where

$$\Delta\varphi = \gamma L_{\mathit{eff}} \left(P_s + 2P_p \right) \qquad (8.14)$$

is the nonlinear phase shift, L_{eff} is the effective interaction length, and $P_j = |U_j|^2$ is the power of the signal ($j = s$) or pump ($j = p$) waves. The first term in Eq. (8.14) is caused by self-phase modulation (SPM), whereas the second term is due to the XPM effect. This term determines the conversion efficiency of the XPM-based wavelength conversion. When Raman gain is present, the accumulated phase shift for a given signal power level is increased. This effect is due to the increased effective interaction length, L_{eff}, which becomes

$$L_{\mathit{eff}} = \frac{1 - \exp\{-(\alpha - g)L\}}{\alpha - g} \qquad (8.15)$$

where g is the gain provided by the Raman amplification.

The Raman gain enhancement of a regenerative ultrafast all-optical XPM WC was experimentally demonstrated in Ref. [57] at 40 and 80 Gb/s. The XPM conversion of a 40 Gb/s OTDM signal was performed with ~1 dB penalty in the receiver sensitivity, while the conversion of a strongly degraded 80 Gb/s signal produced ~2 dB sensitivity improvement, demonstrating signal regeneration. The Raman gain greatly enhanced the wavelength conversion efficiency at 80 Gb/s by 21 dB at a Raman pump power of 600 mW using 1 km of a HNLF.

From Eqs. (8.14) and (8.15), it can be seen that in order to achieve a given phase shift, the signal launch power can be reduced in the presence of Raman gain. On the other hand, for a given CW launch power, more power will be available in the wavelength converted output signal if it is amplified while passing through the fiber. As a consequence, less amplified spontaneous emission noise will be generated when the converted signal is amplified at the output of the converter. In a 2007 experiment, Raman-assisted wavelength conversion by XPM was demonstrated using a 500-m-long HNLF at the record-high bit rate of 640 Gb/s [58]. In this experiment, the noise level after filtering and amplification was reduced by 3.5 dB, while the signal power was increased by ~2 dB.

In a 2006 experiment, wavelength conversion was performed using XPM in a 1-m-long As_2Se_3 fiber, which had a nonlinear parameter $\gamma \sim 1200$ W^{-1}/km and a dispersion $D = -560$ ps/(nm-km) at 1550 nm [59]. The optical data was amplified and a 1.3 nm tunable bandpass filter (TBF) was used to remove out-of-band amplified spontaneous emission (ASE). The pulses were combined with a CW probe from a wavelength tunable amplified laser diode and coupled into the As_2Se_3 fiber. In-line polarization controllers ensured that the polarization state of the pump and probe were aligned. The output of the As_2Se_3 fiber was then sent through a sharp 0.56 nm tunable grating filter offset to longer wavelengths by 0.55–0.70 nm to remove the pump and select a single XPM sideband. A second 1.3 nm TBF was used to remove out-of-band ASE of the amplified signal. An in-line, 200 pm wide, fiber Bragg grating notch filter was also used to further suppress the residual CW. Using such a setup, error-free conversion was demonstrated at 10 Gb/s RZ near 1550 nm over 10nm wavelength range with 7.1 ps pulses at 2.1 W.

XPM-based wavelength conversion was also demonstrated in a 1-m-long bismuth oxide fiber over 15-nm wavelength range at 160 Gb/s [60]. In fact, the As_2Se_3 fiber presents the highest Kerr nonlinearity to date, which is an order of magnitude greater than Bi_2O_3. However, the much larger effective core area of the As_2Se_3 fiber used in [59] compared with that of the Bi_2O_3 fiber in [60] resulted in a similar nonlinear parameter γ in both cases. An effective core area of the As_2Se_3 fiber similar to that of the Bi_2O_3 fiber would result in a nonlinear parameter $\gamma = 11,100$ W^{-1}/km. Such reduction of the As_2Se_3 fiber core area will significantly reduce the device length and fiber losses, as well as dispersion-related impairments.

Pump-probe walk-off is an issue in XPM-based WCs, since it limits the conversion bandwidth. A typical conversion bandwidth of 10 nm can be achieved for a 1-m-long Bi_2O_3 devices. In spite of the fact that the dispersion parameter of As_2Se_3 is higher than that of Bi_2O_3, it is expected that using shorter As_2Se_3 fiber lengths it will be possible to achieve a conversion bandwidth >40 nm. In contrast with silica DSFs, both As_2Se_3 and Bi_2O_3 fibers present a relatively constant dispersion over the whole communications band [61].

A different scheme for wavelength conversion of complex (amplitude and phase) optical signals was proposed and experimentally demonstrated in 2015 using an all-optical implementation of the concept of time-domain holography [62]. In this scheme, wavelength conversion is considered as a modulation process where the complex envelope of the information signal modulates a different carrier signal. The temporal holograms were generated by phase-only modulation, namely XPM, in a HNLF. The scheme inherently provides a wavelength-converted temporal conjugated copy of the original signal additional to the non-conjugated copy. Although the proposed configuration is similar to that of an FWM-based dual pump configuration, it avoids the need to satisfy the phase-matching condition and also relaxes significantly the power requirements [62].

A different type of XPM-based WC consists of the use of a CW pump that is sinusoidally modulated at a frequency f_m higher than the signal bit rate [63]. The pump is amplified and interacts with the signal channel through XPM. As a consequence, the spectrum of signal channel develops amplitude modulation (AM) sidebands separated in frequency by f_m. The use of a suitable optical filter allows the selection of a specific sideband, which represents the wavelength-converted signal.

In a variant of this technique, the sinusoidally modulated pump can be obtained from beating two CW lasers with a small wavelength spacing. Then the data channel will be phase modulated with the beat note frequency through which it will develop the sidebands. The optical frequency separation of the two pump waves determines the amount of optical frequency shift. Different polarization schemes can be implemented for XPM generation. XPM is driven most strongly when the two pump waves are in the parallel polarization state. On the other hand, the phase-modulation index is maximal when the signal is polarized parallel to the pump waves and is minimal when the signal is polarized orthogonal to the pump waves. In the last case, the power of generated signal is lower than that generated by the parallel polarization inputs by a factor of 4 (6 dB) [64–66]. Despite the decrease in conversion efficiency, an advantage of this alignment is to minimize additional FWM generation, which sometimes results in harmful noise to the converted signal. A scheme with two orthogonal pumps is also possible. However, the power-generation efficiency in this case is 6 dB lower than in the parallel polarization case [64–66].

In a 2019 experiment, the performance of the two-pump XPM WC was investigated using a 300-m-long HNLF with a loss coefficient $\alpha \sim 0.5$ dB/km and a nonlinear parameter $\gamma \approx 15\,\text{W}^{-1}\text{km}^{-1}$ [66]. Each CW pump wave was amplified by an Erbium-doped fiber amplifier and phase modulated at a frequency of $f = 50$ MHz to suppress the SBS in the HNLF. It was demonstrated that such scheme is able to generate a phase preserving copy of the optical signal at an exact frequency up-/down-shifted by the two-pump detuning. A conversion bandwidth over 4 THz and error-free wavelength conversion for a 32-GBd polarization-division multiplexed 16QAM signal were demonstrated.

REFERENCES

1. J. Lian, S. Fu, Y. Meng, M. Tang, P. Shum, and D. Liu, *Opt. Express* **22**, 22996 (2014).
2. P. Steinvurzel, J. Demas, B. Tai, Y. Chen, L. Yan, and S. Ramachandran, *Opt. Lett.* **39**, 743 (2014).
3. S. Lavdas, S. Zhao, J. B. Driscoll, R. R. Grote, R. M. Osgood, and N. C. Panoiu, *Opt. Lett.* **39**, 4017 (2014).
4. J. Demas, P. Steinvurzel, B. Tai, L. Rishøj, Y. Chen, and S. Ramachandran, *Optica* **2**, 14 (2015).

5. J. M. Chavez Boggio, J. R. Windmiller, M. Knutzen, R. Jiang, C. Brès, N. Alic, B. Stossel, K. Rottwitt, and S. Radic, *Opt. Express* **16**, 5435 (2008).
6. T. Hasegawa, K. Inoue, and K. Oda, *IEEE Photonics Technol. Lett.* **5**, 947 (1993).
7. G.-W. Lu, T. Sakamoto, and T. Kawanishi, *Opt. Express* **22**, 15 (2014).
8. T. Inoue, K. Tanizawa, and S. Namiki, *J. Lightwave Technol.* **32**, 1981 (2014).
9. H. N. Tan, T. Inoue, T. Kurosu, and S. Namiki, *J. Lightwave Technol.* **34**, 633 (2016).
10. S. Radic and C. J. McKinstrie, *Opt. Fiber Technol.* **9**, 7 (2003).
11. I. Sackey, A. Gajda, A. Peczek, E. Liebig, L. Zimmermann, K. Petermann, and C. Schubert, *Opt. Express* **25**, 21229 (2017).
12. F. Da Ros, M. P. Yankov, E. P. da Silva, M. Pu, L. Ottaviano, H. Hu, E. Semenova, S. Forchhammer, D. Zibar, M. Galili, K. Yvind, and L. K. Oxenløwe, *J. Lightwave Technol.* **35**, 3750 (2017).
13. J. Leuthold, C. H. Joyner, B. Mikkelsen, G. Raybon, J. L. Pleumeekers, B. I. Miler, K. Dreyer, and C. A. Burrus, *Electron. Lett.* **36**, 1129 (2000).
14. M. L. Masanovic, V. Lal, J. S. Barton, E. J. Skogen, L. A. Coldren, and D. J. Blumenthal, *IEEE Photon. Technol. Lett.* **15**, 1117 (2003).
15. A. E. Kelly, D. D. Marcenac, and D. Neset, *Electron. Lett.* **33**, 2123 (1997).
16. M. F. Ferreira, *Nonlinear Effects on Optical Fibers*, John Wiley & Sons, Hoboken, NJ (2011).
17. S. Radic, C. J. McKinstrie, A. R. Chraplyvy, G. Raybon, J. C. Centanni, C. G. Jorgensen, K. Brar, and C. Headley, *IEEE Photonics Technol. Lett.* **14**, 1406 (2002).
18. J. Hansryd, P. A. Andrekson, M. Westlund, J. Li, and P. O. Hedekvist, *IEEE J. Sel. Top. Quantum Electron.* **8**, 506 (2002).
19. J. F. Víctor, F. Rancaño, P. Parmigiani, P. Petropoulos, and D. J. Richardson, *J. Lightwave Technol.* **32**, 3027 (2014).
20. J. Hansryd and P. A. Andrekson, *IEEE Photonics Technol. Lett.* **13**, 194 (2001).
21. J. M. Chavez Boggio, J. R. Windmiller, M. Knutzen, R. Jiang, C. Brès, N. Alic, B. Stossel, K. Rottwitt, and S. Radic, *Opt. Express* **16**, 5435 (2008).
22. M. Hirano, T. Nakanishi, T. Okuno, and M. Onishi, *IEEE J. Sel. Top. Quantum Electron.* **15**, 103 (2009).
23. D. Nodop, C. Jauregui, D. Schimpf, J. Limpert, and A. Tünnermann, *Opt. Lett.* **34**, 3499 (2009).
24. O. Aso, S. Arai, T. Yagi, M. Tadakum, Y. Suzuki, and S. Namiki, *Electron. Lett.* **36**, 709 (2000).
25. G. A. Nowak, Y.-H. Kao, T. J. Xia, M. N. Islam, and D. Nolan, *Opt. Lett.* **23**, 936 (1998).
26. M. N. Islam and O. Boyraz, *IEEE J. Sel. Topics Quantum Electron.* **8**, 527 (2002).
27. S. R. Petersen, T. T. Alkeskjold, and J. Lægsgaard, *Opt. Express* **21**, 18111 (2013).
28. J. Yuan, X. Sang, Q. Wu, G. Zhou, F. Li, X. Zhou, C. Yu, K. Wang, B. Yan, Y. Han, H. Y. Tam, and P. K. A. Wai, *Opt. Lett.* **40**, 1338 (2015).
29. J. Yuan, X. Sang, Q. Wu, G. Zhou, C. Yu, K. Wang, B. Yan, Y. Han, G. Farrell, and L. Hou, *Opt. Lett.* **38**, 5288 (2013).
30. K. M. Hilligsøe, T. Andersen, H. Paulsen, C. Nielsen, K. Mølmer, S. Keiding, R. Kristiansen, K. Hansen, and J. Larsen, *Opt. Express* **12**, 1045 (2004).
31. T. Andersen, K. Hilligsøe, C. Nielsen, J. Thøgersen, K. Hansen, S. Keiding, and J. Larsen, *Opt. Express* **12**, 4113 (2004).
32. X. Zhao, X. Liu, S. Wang, W. Wang, Y. Han, Z. Liu, S. Li, and L. Hou, *Opt. Express* **23**, 27899 (2015).
33. M. Takahashi, R. Sugizaki, J. Hiroishi, M. Tadakuma, Y. Taniguchi, and T. Yagi, *J. Lightwave Technol.* **23**, 3615 (2005).
34. J. M. Chavez Boggio, S. Zlatanovic, F. Gholami, J. M. Aparicio, S. Moro, K. Balch, N. Alic, and S. Radic, *Opt. Express* **18**, 439 (2010).

35. F. Gholami, S. Zlatanovic, E. Myslivets, S. Moro, B. P.-P. Kuo, C.-S. Brès, A. O. J. Wiberg, N. Alic, and S. Radic, *Proc. OFC/NFOEC 2011*, paper OThC6, (2011).
36. F. Gholami, B. P.-P. Kuo, S. Zlatanovic, N. Alic, and S. Radic, *Opt. Express* **21**, 11415 (2013).
37. S. Xing, D. Grassani, S. Kharitonov, A. Billat, and C. Brès, *Opt. Express* **24**, 9741 (2016).
38. M. E. Marhic, P. A. Andrekson, P. Petropoulos, S. Radic, C. Peucheret, and M. Jazayerifar, *Laser Photonics Rev.* **9**, 50 (2015).
39. M. Hirano, T. Nakanishi, T. Okuno, and M. Onishi, *2006 European Conference on Optical Communications*, **4**, 21 (2006).
40. C. J. McKinstrie and S. Radic, *Opt. Lett.* **27**, 1138 (2002).
41. F. Parmigiani, P. Horak, Y. Jung, L. Grüner-Nielsen, T. Geisler, P. Petropoulos, and D. J. Richardson, *Opt. Express* **25**, 33602 (2017).
42. H. Zhang, M. Bigot-Astruc, L. Bigot, P. Sillard, and J. Fatome, *Opt. Express* **27**, 15413 (2019).
43. O. Anjum, M. Guasoni, P. Horak, Y. Jung, M. Suzuki, T. Hasegawa, K. Bottrill, D. Richardson, F. Parmigiani, and P. Petropoulos, *Opt. Express* **27**, 24072 (2019).
44. S. M. M. Friis, I. Begleris, Y. Jung, K. Rottwitt, P. Petropoulos, D. J. Richardson, P. Horak, and F. Parmigiani, *Opt. Express* **24**, 30338 (2016).
45. O. F. Anjum, P. Horak, Y. Jung, M. Suzuki, Y. Yamamoto, T. Hasegawa, P. Petropoulos, D. J. Richardson, and F. Parmigiani, *APL Photonics* **4**, 022902 (2019).
46. R. Elschner, C. A. Bunge, B. Huttl, A. G. Coca, C. S. Langhorst, R. Ludwig, C. Schubert, and K. Petermann, *IEEE J. Sel. Topics Quantum Electron.* **14**, 666 (2008).
47. M.-C. Ho, M. E. Marhic, K. Y. K. Wong, and L. G. Kazovsky, *J. Lightwave Technol.* **20**, 469 (2002).
48. T. Tanemura and K. Kikuchi, *IEEE Photon. Technol. Lett.* **15**, 1573 (2003).
49. X. Guo and C. Shu, *Opt. Express* **24**, 28648 (2016).
50. K. K. Chow, C. Shu, C. Lin, and A. Bjarklev, *IEEE Photon. Technol. Lett.* **17**, 624 (2005).
51. J. H. Lee, K. Kikuchi, T. Nagashima, T. Hasegawa, S. Ohara, and N. Sugimoto, *Opt. Express* **13**, 3144 (2005).
52. M. P. Fok and C. Shu, *IEEE J. Sel. Topics Quantum Electron.* **14**, 587 (2008).
53. J. Yu, X. Zheng, C. Peucheret, A. T. Clausen, H. N. Poulsen, and P. Jeppesen, *J. Lightwave Technol.* **18**, 1001 (2000).
54. B. E. Olsson, P. Öhlén, L. Rau, and D. J. Blumenthal, *IEEE Photon. Technol. Lett.* **12**, 846 (2000).
55. K. Igarashi and K. Kikuchi, *IEEE J. Sel. Topics Quantum Electron.* **14**, 551 (2008).
56. J. Yu and P. Jeppese, *IEEE Photon. Technol. Lett.* **13**, 833 (2001).
57. W. Wang, H. N. Poulsen, L. Rau, H.-F. Chou, J. E. Bowers, and D. J. Blumenthal, *J. Lightwave Technol.* **23**, 1105 (2005).
58. M. Galili, L. K. Oxenlowe, H. C. H. Hansen, A. T. Clausen, and P. Jeppesen, *IEEE J. Sel. Top. Quantum Electron.* **14**, 573 (2008).
59. V. G. Ta'eed, L. B. Fu, M. Pelusi et al., *Opt. Express* **14**, 10371 (2006).
60. J. H. Lee, T. Nagashima, T. Hasegawa et al., *Electron. Lett.* **41**, 918 (2005).
61. L. B. Fu, M. Rochette, V. G. Ta'eed et al. *Opt. Express* **13**, 7637 (2005).
62. M. Fernández-Ruiz, L. Lei, M. Rochette, and J. Azaña, *Opt. Express* **23**, 22847 (2015).
63. W. Mao, P. A. Andrekson, and J. Toulouse, *IEEE Photon. Technol. Lett.* **17**, 42 (2005).
64. C. McKinstrie, S. Radic, and M. Raymer, *Opt. Express* **12**, 5037 (2004).
65. K. Inoue, *IEEE J. Quantum Electron.* **28**, 883 (1992).
66. S. Watanabe, T. Kato, T. Tanimura, C. Schmidt-Langhorst, R. Elschner, I. Sackey, C. Schubert, and T. Hoshida, *Opt. Express* **27**, 16767 (2019).

9 Optical Regeneration

9.1 INTRODUCTION

Optical signals propagating in fiber-optic transmission systems are affected by several effects, namely amplified spontaneous emission (ASE) from optical amplifiers, chromatic dispersion, polarization-mode dispersion, and nonlinear phenomena. Considering the impairments imposed by these effects, the use of signal regenerators becomes necessary in order to extend the maximum transmission distance of the systems. Devices that perform the reamplification, reshaping, and retiming of the degraded bit stream are called 3R regenerators, whereas those able to perform only the first two functions are called 2R regenerators.

The signal regeneration is conventionally performed by optical-electrical-optical (OEO) regenerators that involve signal conversion between the optical and electrical domains and noise removal on the electrical signal. Another type of regeneration that does not rely on the OEO conversion is provided by the all-optical regenerators. Compared with conventional techniques, all-optical signal regeneration offers much higher response speeds, lower power consumption, lower costs, and flexibility in handling signals in different modulation formats [1–3].

One promising medium to perform all-optical signal regeneration is the highly nonlinear fiber (HNLF), made from silica or from other glass materials having much higher nonlinearity [4–10]. As seen in previous chapters, pure and stable nonlinear phase shifts with ultrafast response time can be obtained with these fibers, which lead to well-controlled high-speed operation of signal processing devices. Different types of nonlinear effects can be used for such purpose, such as self-phase modulation (SPM), cross-phase modulation (XPM), four-wave mixing (FWM), and stimulated Raman scattering (SRS) [8–18].

9.2 2R REGENERATORS

All the three major nonlinear effects, SPM, XPM, and FWM, can be used to realize 2R regeneration. Among them, SPM is most often used. An advantage of the SPM-based 2R regenerators is that neither pump nor probe optical sources are needed inside the regenerator, which makes it particularly simple.

9.2.1 SPM-Based Regenerators

2R optical regeneration using an HNLF with subsequent spectral filtering was first proposed by Mamyshev [11]. This scheme relies on SPM to broaden the spectrum of a return-to-zero (RZ) input signal and has been the object of a lot of attention, since it presents a number of key features [19–26]. Among them are the simplicity, a bandwidth that is limited only by the intrinsic material nonlinear response, and the possibility to achieve a direct bit error rate (BER) improvement [25,26]. Actually, a

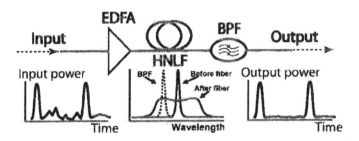

FIGURE 9.1 An SPM-based 2R regenerator (top) and its action on a bit stream (bottom). HNLF stands for a highly nonlinear fiber. (From M. Rochette, et al., *IEEE J. Select. Topics Quantum Electron.*, 12, 736, 2006. © 2006 IEEE. With permission.)

SPM-based regenerator has the important characteristic of discriminating pulses of different widths. This capacity is enabled from the slope dependence of the spectral broadening in SPM, which discriminates logical ones from logical zeros and thereby provides a BER improvement. Up to 1 million kilometers of unrepeated transmission was reported in a 2002 experiment using such regenerator [27].

Figure 9.1 illustrates the operation of a Mamyshev regenerator. A noisy input RZ signal is phase modulated by its own waveform (SPM), while it propagates through a nonlinear fiber. The temporal phase variation induces the instantaneous frequency shift, and the signal spectrum is broadened. The intensity slope point of the original signal is spectrally shifted and can be selected using a bandpass filter (BPF) offset from the input center wavelength. Furthermore, above a certain level, any variation in the signal intensity will translate into a change in the width of the SPM-induced broadened spectrum. Since the bandwidth of the optical filter is fixed, the intensity of the filtered output is insensitive to those variations of the signal power, which results in an effective suppression of the amplitude noise at bit "one" level. On the other hand, optical noise at bit "zero" is low and does not induce any spectral shift, being filtered out by the BPF. This results in an improvement of the optical signal-to-noise ratio (SNR).

The SPM-based signal processing can be characterized by three parameters: the instantaneous frequency shift induced by phase modulation, f_{PM}, the BPF offset frequency relative to the original signal center frequency, f_{BPF}, and the BPF bandwidth, Δf_{BPF}. The requirement for f_{PM} in order to guarantee a proper operation is given by [28]:

$$f_{PM} > 2\Delta f_{in} \tag{9.1}$$

where Δf_{in} is the bandwidth of the incoming optical signal. The maximum value of SPM-induced frequency shift depends on the pulse shape. Assuming that Gaussian pulses with a peak power P_0 are launched into an optical fiber of length L, Eq. (9.1) provides the following condition [28]:

$$\gamma P_0 L > \frac{4\pi}{3.45} \tag{9.2}$$

where γ is the nonlinear parameter of the fiber. Considering a 500-m-long silica-based HNLF, such that $\gamma = 20$ W^{-1}/km, the required peak power is $P_0 \approx 0.4$ W.

2R regenerators based on SPM with subsequent spectral filtering have been demonstrated using different types of HNLFs [8,9,20–30]. Since logical "ones" (e.g., pulses) and logical "zeros" (e.g., ASE noise) are processed with distinct transfer functions, such regenerators are able to directly improve the BER of a noisy signal. In a 2006 experiment, a BER improvement from 5×10^{-9} to 4×10^{-12} for an optical SNR ratio of 23 dB was achieved using a 851-m-long highly nonlinear silica fiber [26]. A BPF with a bandwidth of 0.56 nm was offset 1.2 nm from the signal center wavelength. A more compact version of such 2R regenerator was realized in the same experiment using a 2.8-m-long highly nonlinear As_2Se_3 chalcogenide fiber. Chalcogenide glass offers many advantages, such as its large Kerr nonlinearity (up to 1000 × silica glass) and a short response time <100 fs.

Taking into account the two-photon absorption (TPA) effect was essential to describe properly the experimental results of the preceding experiment. The TPA affects the shape and contributes to flattening the upper level of the step-like power transfer function. The corresponding curves exhibited a threshold at 1 W and an output power limiting function at ~8 W peak input power. The first feature was important to suppress the noise in the "zeros," whereas the second feature helped in suppressing the noise in the "ones."

The As_2Se_3 fiber used in the regenerator had a linear normal dispersion of 504 ps/nm/km at 1550 nm. This large value of the dispersion proves to be important in smoothing out the nonlinear transfer curves at high-peak power levels [26]. The total dispersion of 1–2 ps/nm present in the device is similar to that of a typical silica HNLF with one to several kilometers used in some regeneration experiments [21]. However, in the case of the As_2Se_3 fiber, such level of dispersion corresponds to a very short length of a few meters. Besides the advantages resulting from the short length, a reduction of the operating powers by over an order of magnitude is expected by reducing the mode field diameter.

In a 2008 experiment, 2R regeneration of an amplitude shifted keying data signal was also achieved using a 35-cm long highly nonlinear bismuth oxide fiber (Bi-NLF) [29]. In this experiment, a fiber laser was harmonically mode-locked at 10 GHz at a center wavelength of 1550.66 nm. The output was externally modulated using an electrooptic modulator, which was biased in order to intentionally degrade the signal extinction ratio. The degraded signal was then amplified with an erbium-doped fiber amplifier (EDFA) and afterward launched to the Bi-NLF. The measured eye diagram of the distorted 10-Gb/s RZ input signal presented an extinction ratio of only 7.4 dB. After undergoing SPM in the Bi-NLF, the signal degradation still remained. However, using a BPF placed 2 nm away from the spectral peak of the input signal, a regenerated signal was achieved. The extinction ratio of the regenerated signal was improved to 15.5 dB. The BER measurement results of the RZ data signal have shown a sensitivity improvement of 5 dB at a 10^{-9} BER [29].

SPM-based 2R regenerators can also be used when soliton-like pulses propagate in the anomalous dispersion regime of an optical fiber. In such case, the input pulse is compressed or decompressed, depending on whether its amplitude is larger or smaller than that of the fundamental soliton in the fiber. As a consequence, the spectral width of the pulse is broadened or narrowed, respectively. By using a bandpass optical filter at the fiber output, larger (smaller) energy loss is induced to the pulse having larger (smaller) initial amplitude, which provides a mechanism of self-amplitude stabilization.

To compensate for the filter-induced loss, some excess gain must be provided to the soliton. However, this excess gain also amplifies the small-amplitude waves coexistent with soliton. Such amplification results in a background instability, which can significantly affect and even destroy the soliton itself, especially when a number of regenerators are cascaded. An approach to avoid the background instability consists of using filters whose central frequency is gradually shifted along the cascaded regenerators. This is similar to the soliton control technique based on the use of sliding-frequency guiding filters in long-distance transmission [31]. Another way to suppress the noise growth is to use saturable absorbing elements [32,33]. This technique has been used in a 2002 experiment, in which soliton pulses were transmitted at 40 Gb/s along 7600 km, with a regenerator spacing of 240 km [34].

It can be seen from Eq. (8.11) that the transmittivity of a nonlinear optical loop mirror (NOLM) has a minimum value at vanishing input power, given by

$$T_{\min} = 1 - 4f(1-f) \tag{9.3}$$

where f is the coupler power-splitting fraction. This shows that the NOLM can be used to suppress the zero-level noise, which becomes more effective for a coupling ratio near 0.5 [35,36]. On the other hand, the NOLM can also suppress the amplitude fluctuations of the input signal if its power is set at one of the values at which the output power is locally constant. In a 2004 experiment, 42.66 Gb/s signal transmission over 10,000 km with a Q factor of 11 dB in a five-channel wavelength division multiplexing system was demonstrated with NOLM-based 2R regenerators inserted every 240 km [37]. Noise and pedestal suppression have been also demonstrated using the nonlinear amplifying loop mirror (NALM) mechanism [36,38,39].

9.2.2 SPM-Based Optical Pulse Train Generation

Besides the regeneration function described in the previous section, the SPM-based pulse reshaping in HNLFs can also be used with advantage in the generation of optical pulse trains. This application is particularly important for 160-Gb/s optical time-division multiplexing (OTDM) transmitters, where the generation of 10-GHz picosecond optical pulse trains is one main issue. High-quality pulse trains are required for such OTDM transmitters, namely a reduced pulse width, a low timing jitter, and a long-term stability, as well as wavelength and repetition-rate tunability.

Using this mechanism, a pulse-train generator with a repetition rate of 10 GHz was realized in a 2008 experiment [28]. First, a continuous wave (CW) light was sinusoidally phase modulated by an $LiNbO_3$ phase modulator driven by a 10-GHz clock. The chirp induced by the phase modulation was compensated by a dispersion compensating fiber as a dispersive medium. The phase-modulated wave was then converted to a picosecond pulse train, which was accompanied with a large pedestal. Such a pedestal was eliminated by the SPM-based pulse reshaping in an optical fiber [11]. Since the spectrum of the pedestal remains in the vicinity of the center wavelength of the input spectrum, it was suppressed using a BPF offset from this frequency. Finally, a standard single-mode fiber was used to compensate for the residual chirp of the reshaped pulse. Highly stable generation of a 10-GHz 2-ps optical pulse train tunable over the entire C band was achieved using such a pulse generator, which incorporated a 500-m-long highly nonlinear silica fiber. The applicability of such a pulse generator to the 160-Gb/s transmitter has been also demonstrated [28].

9.2.3 Femtosecond Pulse Generation

During the last few years, the principle of Mamyshev regenerator has found interesting applications in the area of femtosecond pulse generation. For example, it allows generation of femtosecond pulses using picosecond Q-switched laser [40] or gain-switched laser [41]; it has also been applied to enhance the temporal contrast of mJ femtosecond pulses [42]. Actually, a new kind of mode-locked lasers has been implemented making use of a Mamyshev regenerator inside the laser cavity [43,44].

Wavelength widely tunable (~400 nm) femtosecond sources have been achieved using the Mamyshev regenerator idea previously described [45–47]. In this case, the SPM-dominated nonlinearity is employed to dramatically broaden an input narrowband optical spectrum to a spectral width of 100 s of nm. Using optical filters to select the leftmost/rightmost spectral lobes of the broadened spectrum, nearly transform-limited ~100-fs pulses with the center wavelength widely tunable can be obtained. In a 2018 experiment, an ultrafast Er-fiber laser and a 7-cm long dispersion-shifted fiber with negative group-velocity dispersion (GVD) were used to implement a tunable source that produced ~100-fs pulses tunable between 1.3 and 1.7 µm with up to ~16-nJ pulse energy [48].

9.2.4 FWM-Based 2R Regenerators

As seen in Section 2.5, when a signal at frequency f_s is launched into a fiber-optic parametric amplifier (FOPA) together with a pump at frequency f_p, the FWM process amplifies the signal and generates an idler wave with frequency $f_i = 2f_p - f_s$. As the input signal power is increased, the direction of power flow from pump to the signal can be reversed, and higher-order FWM products are generated [49–51]. These lead to gain saturation of the parametric amplifier, which can be used to reduce the fluctuations in the signal pulse peak power [52,53]. An important feature of this type of FWM-based regenerator is that the optical carrier phase of the input signal is basically preserved [54], which indicates that it can be used for regeneration of phase-shift keying (PSK) signals. A basic single-pump FOPA consists of a pump source, a HNLF, and an optical BPF for extracting the output signal wave.

The idler wave and the higher-order FWM products can also be used to realize simultaneously the 2R regeneration and the wavelength conversion of an input signal. The output powers of the higher-order FWM components at $f_2 = 2f_s - f_p$ and $f_3 = 2f_s - f_i$ grow proportionally to the square and cube of the input signal power, respectively, in the small signal regime [55]. This means that the low power noise is more effectively suppressed using these components. On the other hand, the saturation that these waves also experience at higher input powers provides the mechanism for pulse amplitude equalization and the realization of the 2R function [56–59]. An issue in using higher-order FWM products is that the input signal phase is not preserved, which precludes the use of this regeneration scheme in the case of PSK signals.

9.3 3R REGENERATORS

In spite of all the advantages offered by 2R regeneration, it is not able to retime the signal and therefore cannot reduce the timing jitter. Full regeneration (3R) is accomplished when the retiming function is added to reamplification and reshaping of the optical signals. An SPM-based 3R regenerator can be realized by adding a synchronous amplitude modulator to the scheme shown in Figure 9.1. An electrical clock signal recovered from the incoming bit stream itself is used to drive the modulator. In a 2002 experiment, this approach has been used to transmit a 40-Gb/s signal over one million km [60]. The same technique has been used also in the context of soliton systems to suppress the background instability [61].

Full regeneration can also be achieved using XPM [14,62,63]. The configuration of XPM-based 3R regenerator was first proposed by Suzuki et al. [14] and consists basically of an optical clock source, an EDFA, an HNLF, and a BPF, as illustrated in Figure 9.2. An impaired signal with a center wavelength λ_1 is strongly pumped by an EDFA. Such signal is launched into a highly nonlinear dispersion-shifted fiber (HNL-DSF) together with a clean clock pulse train at λ_2. A BPF with a center wavelength λ_2 is placed behind the HNL-DSF. The clock is generated from a clock-recovery circuit and produces pulses of shorter duration than the pulses to regenerate. The repetition

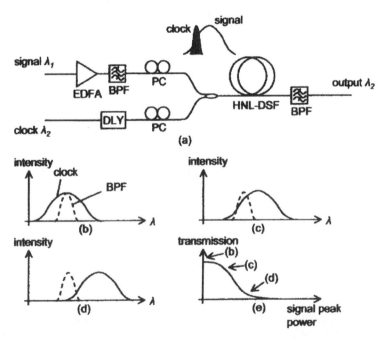

FIGURE 9.2 (a) Schematic and principle of the 3R regenerator based on XPM-induced spectral shift followed by filtering. The peak power of the signal pulse is increased from (b) to (d). (From J. Suzuki, et al., *IEEE Photon. Technol. Lett.*, 17, 423, 2005. © 2005 IEEE. With permission.)

rate of the clock can be chosen equal to the signal data rate when regeneration is desired. However, the repetition rate of the clock can be smaller than the signal data rate when regeneration and demultiplexing are desired. As the signal bits and clock pulses temporally overlap in the HNLF, the clock spectrum experiences an XPM-induced frequency shift proportional to the slope of the superimposed signal, which depends on the peak power, as represented in Figure 9.2b–d. The output power of the clock pulse from the BPF is then a nonlinear function of the signal peak power, as shown in Figure 9.2e. The output filter BPF filters out the spectral components that have been shifted by XPM and keeps the original frequency components at wavelength λ_2. Note that this regeneration scheme inverts the bits, that is, logical ones are transformed into logical zeros and vice versa. A noninverting mode can also be achieved by offsetting the center frequency of output BPF away from the original clock wavelength, in order to transmit only XPM generated frequencies.

A common challenge of nonlinear effects such as FWM and XPM is their inherent sensitivity to the state-of-polarization (SOP) of the input signal. Since the SOP of a data signal varies randomly with time in practical systems, it is important that the all-optical retiming function be polarization independent. Several techniques have been proposed to address this problem, namely, depolarization of the clock signal [64], polarization diversity [65], fiber twisting for circularly polarizing fiber [66], and the use of fiber birefringence [67]. However, these techniques often require additional complexity and/or careful control and stability of the clock signal SOP.

Polarization-independent all-optical retiming without additional complexity has been demonstrated using XPM in an HNLF and offset spectral filtering [17,68]. This technique is based on the fact that there exist wavelength regions for which the XPM-broadened spectrum is polarization independent. An all-optical 3R regenerator using such an approach has been demonstrated in a 2008 experiment [17]. The regenerator used a self-pulsating distributed feedback (DFB) laser for clock recovery, XPM in an HNLF and offset spectral slicing for retiming, SPM in an HNLF and offset spectral slicing for reshaping. In the retiming stage, an optical clock signal was recovered and used to induce XPM on the input data signal in an HNLF, followed by offset spectral filtering. A tunable optical delay was implemented to control the temporal alignment of the clock and data signals at the input of the HNLF. The retimed signal was obtained by filtering the broadened spectrum of the data signal with a BPF offset from the carrier wavelength of the data signal by $\Delta\lambda$. The subsequent reshaping stage used SPM in an HNLF and spectral filtering with an offset of $-\Delta\lambda$, thus preserving the signal wavelength. Due to the ultrafast properties of the Kerr effect in HNLF, the proposed technique can be applied to 40 and 80 Gb/s RZ signals with the use of a polarization-insensitive 40 or 80 GHz self-pulsating laser [17].

9.4 ALL-OPTICAL REGENERATION OF PHASE-ENCODED SIGNALS

Increasing needs for bandwidth and power-efficient data transmission in long-distance lightwave systems have recently motivated a change of modulation formats from on-off keying (OOK) to PSK and more advanced formats. As a consequence, several schemes for all-optical regeneration of PSK signals have been recently proposed and demonstrated.

In comparison with the OOK format, the differential binary PSK (DPSK) format offers a better sensitivity and better tolerance to fiber nonlinearity. In one class of DPSK regenerators, the phase information of the signal is converted into the amplitude information through the use of a delay interferometer. The phase noise of the signal is thus transferred to the amplitude of the demodulated OOK signal, whose noise is afterward removed by an amplitude regenerator [69–72]. The regenerated OOK signal is used as a control signal to modulate the phase of probe pulses, thus providing a regenerated PSK signal.

Phase-preserving amplitude-only regeneration has also been shown to be effective in suppressing the nonlinear phase-noise [73–76]. This is induced by the in-phase component of the ASE relative to the signal (amplitude noise) that is converted to phase noise via the effect of SPM in the transmission fiber. The nonlinear phase noise dominates over the linear phase noise in long-distance systems. It can be controlled by suppression of the amplitude noise in the system, which can be realized by inserting an amplitude limiter after the transmitter and at every amplifier span.

Phase-preserving amplitude regeneration can be achieved by fiber-optic parametric amplification operating in the saturated regime, as discussed in Section 8.4. The use of a modified NOLM, in which a directional attenuator is inserted in the loop, has also been suggested for the same purpose [73]. A suitable choice of parameters provides a flat phase response at the input signal power at which the output signal power levels off. The use of distributed Raman amplification in the fiber loop has been shown also to make the NOLM phase preserving [77]. In a 2009 experiment, 10 Gbit/s DPSK signal transmission has been demonstrated using NOLM-based amplitude regenerators [75].

In another type of DPSK regenerator, phase-sensitive amplifiers (PSAs) are used to directly suppress the phase noise [78–81]. In this type of regenerator, with simultaneous use of gain saturation, reduction of phase and amplitude noise can be achieved. In a PSA, the in-phase component of the input signal is amplified, whereas the quadrature component is deamplified. As a result, the output phase is squeezed and becomes much smaller than the input phase. In practice, a PSA can be realized by first generating a phase-conjugate copy of the input signal and afterward adding coherently the signal and its phase-conjugate copy. This can be done using a partially degenerate FWM process, in which two pumps are located symmetrically around the signal, thus generating an idler at the same frequency as the input signal $\left(\omega_i = \omega_1 + \omega_2 - \omega_s = \omega_s\right)$. If the phases of the two CW pumps are locked to the carrier frequency of the signal, the generated idler will add coherently to the input signal. In one experiment, phase-locked pumps for the dual-pump degenerate PSA were generated by using a data-pumped FWM process in a fiber for the data-phase erasure followed by injection locking of a semiconductor laser [80].

REFERENCES

1. M. Jinno and M. Abe, *Electron. Lett.* **28**, 1350 (1992).
2. J. K. Lucek and K. Smith, *Opt. Lett.* **18**, 1226 (1993).
3. O. Leclerc, B. Lavigne, D. Chiaroni, and E. Desurvire, *Optical Fiber Telecommunications IVA, Components*; I. Kaminow and T. Li, Eds., Academic, San Francisco, CA, 732 (2002).
4. M. J. Li, S. Li, and D. A. Nolan, *J. Lightw. Technol.* **23**, 3606 (2005).

5. M. Takahashi, R. Sugizaki, J. Hiroishi, M. Tadakuma, Y. Taniguchi, and T. Yagi, *J. Lightwave. Technol.* **23**, 3615 (2005).
6. M. Hirano, T. Nakanishi, T. Okuno, and M. Onishi, *IEEE J. Sel. Topics Quantum Electron.* **15**, 103 (2009).
7. J. H. Lee, P. C. The, Z. Yusoff, M. Ibsen, W. Belardi, T. M. Monro, and D. J. Richardson, *IEEE Photon. Technol. Lett.* **14**, 876 (2002).
8. F. Parmigiani, S. Asimakis, N. Sugimoto, F. Koizumi, P. Petropoulos, and D. J. Richardson, *Opt. Express* **14**, 5038 (2006).
9. L. B. Fu, M. Rochette, V. G. Ta'eed, D. J. Moss, and B. J. Eggleton, *Opt. Express* **13**, 7637 (2005).
10. M. F. Ferreira, *Nonlinear Effects in Optical Fibers*; John Wiley & Sons, Hoboken, NJ (2011).
11. P. V. Mamyshev, *Eur. Conf. Opt. Commun.* (ECOC98), Madrid, Spain, pp. 475–477 (1998).
12. W. A. Pender, P. J. Watkinson, E. J. Greer, and A. D. Ellis, *Electron. Lett.* **31**, 1587 (1995).
13. W. A. Pender, T. Widdowson, and A. D. Ellis, *Electron. Lett.* **32**, 567 (1996).
14. J. Suzuki, T. Tanemura, K. Taira, Y. Ozaki, and K. Kikuchi, *IEEE Photon. Technol. Lett.* **17**, 423 (2005).
15. A. Bogoni, P. Ghelfi, M. Scaffardi, and L. Pot, *IEEE J. Sel. Topics Quantum Electron.* **10**, 192 (2004).
16. J. Yu and P. Jeppesen, *J. Lightw. Technol.* **19**, 941 (2001).
17. C. Ito and J. C. Cartledge, *IEEE J. Sel. Top. Quantum Electron.* **14**, 616 (2008).
18. D. Dahan, R. Alizon, A. Bilenca, and G. Eisenstein, *Electron. Lett.* **39**, 307 (2003).
19. M. Matsumoto, *IEEE Photon. Technol. Lett.* **14**, 319 (2002).
20. M. Matsumoto, *J. Lightw. Technol.* **23**, 1472 (2004).
21. T. Her, G. Raybon, and C. Headley, *IEEE Photon. Technol. Lett.* **16**, 200 (2004).
22. T. Miyazaki and F. Kubota, *IEEE Photon. Technol. Lett,* **16**, 1909 (2004).
23. P. Johannisson and M. Karlsson, *IEEE Photon. Technol. Lett.* **17**, 2667 (2005).
24. A. G. Striegler and B. Schmauss, *J. Lightw. Technol.* **24**, 2835 (2006).
25. M. Rochette, J. N. Kutz, J. L. Blows, D. Moss, J. T. Mok, and B. J. Eggleton, *IEEE Photon. Technol. Lett.* **17**, 908 (2005).
26. M. Rochette, F. Libin, V. Ta'eed, D. J. Moss, and B. J. Eggleton, *IEEE J. Select. Topics Quantum Electron.* **12**, 736 (2006).
27. G. Raybon, Y. Su, J. Leuthold, R. J. Essiambre, T. Her, C. Joergensen, P. Steinvurzel, K. Dreyer, and K. Feder, *Opt. Fib. Commun. Conf.* (OFC2002), Anaheim, CA, PD10-1 (2002).
28. K. Igarashi and K. Kikuchi, *IEEE J. Sel. Topics Quantum Electron.* **14**, 551 (2008).
29. M. P. Fok and C. Shu, *IEEE J. Sel. Topics Quantum Electron.* **14**, 587 (2008).
30. S. Radic, D. J. Moss, and B. J. Eggleton, *Optical Fiber Telecommunications VA, Components and Subsystems*, Academic Press, San Diego, CA. (2008).
31. L. F. Mollenauer, J. P. Gordon, and S. G. Evangelides, *Opt. Lett.* **17**, 1575 (1992).
32. M. Matsumoto and O. Leclerc, *Electron. Lett.* **38**, 576 (2002).
33. M. Gay, M. Costa e Silva, T. N. Nguyen, L. Bramerie, T. Chartier, M. Joindot, J.-C. Simon, J. Fatome, C. Finot, and J.-L. Oudar, *IEEE Photon. Technol. Lett.* **22**, 158 (2010).
34. D. Rouvillain, F. Seguineau, L. Pierre, P. Brindel, H. Choumane, G. Aubin, J.-L. Oudar, and O. Leclerc, *Proc. Opt. Fiber Commun. Conf.*, Paper FD11 (2002).
35. K. Smith, N. J. Doran, and P. G. J. Wigley, *Opt. Lett.* **15**, 1294 (1990).
36. E. Yamada and M. Nakazawa, *IEEE J. Quantum Electron.* **30**, 1842 (1994).
37. F. Seguineau, B. Lavigne, D. Rouvillain, P. Brindel, L. Pierre, and O. Leclerc, *Proc. Opt. Fiber Commun. Conf.*, Paper WN4 (2004).

38. B.-E. Olsson and P. A. Andrekson, *J. Lightw. Technol.* **13**, 213 (1995).
39. M. D. Pelusi, Y. Matsui, and A. Suzuki, *IEEE J. Quantum Electron.* **35**, 867 (1999).
40. R. Lehneis, A. Steinmetz, J. Limpert, and A. Tünnermann, *Opt. Lett.* **39**, 5806 (2014).
41. W. Fu, L. G. Wright, and F. W. Wise, *Optica* **4**, 831 (2017).
42. J. Buldt, M. Müller, R. Klas, T. Eidam, J. Limpert, and A. Tünnermann, *Opt. Lett.* **42**, 3761 (2017).
43. K. Regelskis, J. Želudevičius, K. Viskontas, and G. Račiukaitis, *Opt. Lett.* **40**, 5255 (2015).
44. Z. Liu, Z. M. Ziegler, L. G. Wright, and F. W. Wise, *Optica* **4**, 649 (2017).
45. Z. Liu, C. Li, Z. Zhang, F. X. Kärtner, and G. Chang, *Opt. Express* **24**, 15328 (2016).
46. W. Liu, S.-H. Chia, H.-Y. Chung, R. Greinert, F. X. Kärtner, and G. Chang, *Opt. Express* **25**, 6822 (2017).
47. H.-Y. Chung, W. Liu, Q. Cao, F. X. Kärtner, and G. Chang, *Opt. Express* **25**, 15760 (2017).
48. H. Chung, W. Liu, Q. Cao, L. Song, F. X. Kärtner, and G. Chang, *Opt. Express* **26**, 3684 (2018).
49. G. Cappellini and S. Trillo, *J. Opt. Soc. Am. B* **8**, 824 (1991).
50. K. Inoue and T. Mukai, *Opt. Lett.* **26**, 10 (2001)
51. J. P. Cetina, A. Kumpera, M. Karlsson, and P. A. Andrekson, *Opt. Express* **23**, 33426 (2015)
52. K. Inoue, *Electron. Lett.* **36**, 1016 (2000).
53. Y. Su, L. Wang, A. Agrawal, and P. Kumar, *Electron. Lett.* **36**, 1103 (2000).
54. M. Matsumoto, *IEEE Photon. Technol. Lett.* **17**, 1055 (2005).
55. M. Matsumoto, *IEEE J. Select. Topics Quantum Electron.* **18**, 738 (2012).
56. E. Ciaramella and S. Trillo, *IEEE Photon. Technol. Lett.* **12**, 849 (2000).
57. E. Ciaramella, F. Curti, and S. Trillo, *IEEE Photon. Technol. Lett.* **13**, 142 (2001).
58. K. Inoue, *IEEE Photon. Technol. Lett.* **13**, 338 (2001).
59. S. Radic, C. J. McKinstrie, R. M. Jopson, J. C. Centanni, and A. R. Chraplyvy, *IEEE Photon. Technol. Lett.* **15**, 957 (2003).
60. J. Leuthold, G. Raybon, Y. Su et al., *Electron. Lett.* **38**, 890 (2002).
61. A. Sahara, T. Inui, T. Komukai, H. Kubota, and M. Nakazawa, *J. Lightwave Technol.* **18**, 1133 (2000).
62. M. Rochette, J. L. Blows, and B. J. Eggleton, *Opt. Express* **14**, 6414 (2006).
63. M. Daikoku, N. Yoshikane, T. Otani, and H. Tanaka, *J. Light. Technol.* **24**, 1142 (2006).
64. T. Yang, C. Shu, and C. Lin, *Opt. Express* **13**, 5409 (2005).
65. K. K. Chow, C. Shu, C. Lin, and A. Bjarklev, *IEEE Photon. Technol. Lett.* **17**, 624 (2005).
66. T. Tanemura, J. Suzuki, K. Katoh, and K. Kikuchi, *IEEE Photon. Technol. Lett.* **17**, 1052 (2005).
67. A. S. Leniham, R. Salem, T. E. Murphy, and G. M. Carter, *IEEE Photon. Technol. Lett.* **18**, 1329 (2006).
68. C. Ito and J. C. Cartledge, *IEEE Photon. Technol. Lett.* **20**, 425 (2008).
69. P. Vorreau, A. Marculescu, J. Wang et al., *IEEE Photon. Technol. Lett.* **18**, 1970 (2006).
70. M. Matsumoto and H. Sakaguchi, *Opt. Express* **16**, 11169 (2008).
71. M. Matsumoto and Y. Morioka, *Opt. Express* **17**, 6913 (2009).
72. C. Kouloumentas, M. Bougioukos, A. Maziotis, and H. Avramopoulos, *IEEE Photon. Technol. Lett.* **22**, 1187 (2010).
73. A. G. Striegler, M. Meissner, K. Cvecek, K. Spnsel, G. Leuchs, and B. Schmauss, *IEEE Photon. Technol. Lett.* **17**, 639 (2005).
74. M. Matsumoto and K. Sanuki, *Opt. Express* **15**, 8094 (2007).

75. C. Stephan, K. Sponsel, G. Onishchukov, B. Schmauss, and G. Leuchs, *IEEE Photon. Technol. Lett.* **21**, 1864 (2009).
76. Q. T. Le, L. Bramerie, H. T. Nguyen, M. Gay, S. Lobo, M. Joindot, J.-L. Oudar, and J.-C. Simon, *IEEE Photon. Technol. Lett.* **22**, 887 (2010)
77. S. Boscolo, R. Bhamber, and S. K. Turitsyn, *IEEE J. Quantum Electron.* **42**, 619 (2006).
78. A. Bogris and D. Syvridis, *IEEE Photon. Technol. Lett.* **18**, 2144 (2006).
79. K. Croussore and G. Li, *IEEE J. Sel. Topics Quantum Electron.* **14**, 648 (2008).
80. R. Slavík, F. Parmigiani, J. Kakande et al., *Nat. Photon.* **4**, 690 (2010).
81. Z. Zheng, L. An, Z. Li, X. Zhao, and X. Liu, *Opt. Commun.* **281**, 2755 (2008).

Index

B
bismuth-oxide fiber, 50

C
chalcogenide fiber, 51
Cherenkov radiation, 52
complex Ginzburg-Landau equation, 33
continuous wave, 61
cross-phase modulation, 10

D
dispersion distance, 26
dispersion-managed soliton, 35–37
dispersion-shifted fiber, 105, 110
dispersive waves, 51–53
dissipative soliton, 33–35
    resonance, 34

E
effective fiber length, 17
effective mismatch parameter, 14
endlessly single-mode fiber, 48
erupting soliton, 34

F
femtosecond pulse, 62
fiber Brillouin amplifier, 78–82
fiber loss, 28
fiber nonlinear parameter, 8, 49
fiber parametric amplifier, 82–86, 99–101
fiber Raman amplifier, 73–78
    distributed, 77
    few-mode, 78, 107
    higher-order, 78
    lumped, 77
fission distance, 52
four-wave mixing, 11–15, 53
    Bragg-scattering type, 64, 107
    intermodal, 107
frequency filters, 31
fundamental soliton, 27

G
Gordon-Haus effect, 25, 30
group velocity dispersion, 7

H
higher-order soliton, 28, 38
highly nonlinear silica fiber, 43, 106

I
intrapulse Raman scattering, 29

K
Kerr effect, 5

L
lead-silicate glass fiber, 50

M
Mach-Zehnder interferometer, 91
microstructured fiber, 47–49
modulation instability, 59

N
NALM, 94
NOLM, 39, 93
nonlinear distance, 26
nonlinear gain, 32
nonlinear Schrödinger equation, 6–8
non-silica fibers, 50

O
optical regeneration, 117–124
optical switching, 91–102
    SPM-induced, 91–94
    using FWM, 98–101
    XPM-induced, 95–97
optical time division multiplexing, 95

P
phase-encoded signals, 123
phase sensitive amplifier, 100
picosecond pulses, 59

S
Sagnac interferometer, 92
self-phase modulation, 8

slow light, 82
soliton-effect compression, 38–39
soliton fission, 51–53
soliton order, 27, 65
soliton self-frequency shift, 30, 52
soliton switching, 94
stimulated Brillouin scattering, 18–20
stimulated Raman scattering, 15–18
supercontinuum coherence, 67
supercontinuum generation, 59
    modeling, 64

T

tapered fiber, 44
tellurite glass fiber, 50
third-order dispersion, 29
3R regenerators, 122

timing jitter, 31
2R regenerators, 117–12
    femtosecond pulse generation, 121
    FWM-based, 121
    optical pulse train generation, 120
    SPM-based, 117–120

W

wavelength conversion, 103–113
    FWM-based, 103–109
    XPM-based, 109–113
WDM transmission systems, 68

Z

XPM-induced frequency shift, 11